T0292244

Common Grasses, Legumes and Forbs of the Eastern United States

Common Grasses, Legumes and Forbs of the Eastern United States

Identification and Adaptation

A. Ozzie Abaye
Virginia Polytechnic Institute and State University, Blacksburg, VA, USA

With first edition assistance from James T. Green, Jr. and Edward B. Rayburn.

First edition edited by Douglas S. Chamblee and Thomas Wieboldt

ACADEMIC PRESS
An imprint of Elsevier

Academic Press is an imprint of Elsevier
125 London Wall, London EC2Y 5AS, United Kingdom
525 B Street, Suite 1650, San Diego, CA 92101, United States
50 Hampshire Street, 5th Floor, Cambridge, MA 02139, United States
The Boulevard, Langford Lane, Kidlington, Oxford OX5 1GB, United Kingdom

Notices
Knowledge and best practice in this field are constantly changing. As new research and experience broaden our
understanding, changes in research methods, professional practices, or medical treatment may become necessary.

Practitioners and researchers must always rely on their own experience and knowledge in evaluating and using any
information, methods, compounds, or experiments described herein. In using such information or methods they should
be mindful of their own safety and the safety of others, including parties for whom they have a professional
responsibility.

To the fullest extent of the law, neither the Publisher nor the authors, contributors, or editors, assume any liability for
any injury and/or damage to persons or property as a matter of products liability, negligence or otherwise, or from any
use or operation of any methods, products, instructions, or ideas contained in the material herein.

British Library Cataloguing-in-Publication Data
A catalogue record for this book is available from the British Library

Library of Congress Cataloging-in-Publication Data
A catalog record for this book is available from the Library of Congress

ISBN: 978-0-12-813951-6

For Information on all Academic Press publications
visit our website at https://www.elsevier.com/books-and-journals

  Working together
to grow libraries in
developing countries

www.elsevier.com • www.bookaid.org

Publisher: Charlotte Cockle
Acquisition Editor: Nancy Maragioglio
Editorial Project Manager: Carly Demetre
Production Project Manager: Paul Prasad Chandramohan
Cover Designer: Mark Rogers

Typeset by MPS Limited, Chennai, India

Dedication

This book is dedicated to my adoptive parents, Mr. and Mrs. Samuel W. Morris. From the time I arrived on their farm in Pennsylvania in 1978, they had a much bigger dream for me than I had for myself. They supported every move I made and suggested more. I would also like to acknowledge my biological father Mr. Abaye Metekia. He taught me that love is a key that opens all doors and to love oneself is to love all. The completion of this book would not have been possible without the encouragement and gentle pushes, e.g., "when are you finishing the book?", from my sister, Eleanor Morris Illoway.

—Ozzie Abaye, October 2010

Mr. and Mrs. Samuel W. Morris

Dedication

Contents

Acknowledgments and Contributors

The authors express gratitude to the following institutions and individuals for providing plant images, data, suggestions, editing, and encouragement.

Acknowledgments
The O.M. Scott and Sons Company is thanked for permission to use drawings from *Scott's Guide to the Identification of Grasses*.
The legume sketches are modified from *Legume Culture and Picture Identification: Seedling to Maturity*, Herbert B. Hartwig, Agronomy Department, Cornell University, Ithaca, NY.
Justice Family Farms, particularly Mr. James Justice III, is thanked for financial support.
Special thanks to current and former graduate students Christina Newman, Katie Hurder, Elizabeth Yarber, Pepper Rains, Jonathan Rotz, Matt Webb, Meriem El Hadj, Jennifer Fincham, Ginny Pitman Barnes, and James Daniels and Dr. Chester Foy for helping with many aspects of the book: making seed images, editing, and helping organize the images.
Appreciation is extended to mentors Vivien Gore Allen, Gabriella Varga, David Parrish, Mark Alley, and Steven Hodges.

Image Contributors
James Atland—Oregon State University, North Willamette Research and Extension Center (NWREC), Aurora, OR, United States.
Wynse Brooks—Research Associate, Department of Crop and Soil Environmental Sciences, Virginia Tech, Blacksburg, VA, United States.
Cathi Bonin—Graduate student, Crop and Soil Environmental Sciences, Virginia Tech, Blacksburg, VA, United States.
Lachlan M.D. Cranswick—Deep River, ON, Canada.
Sam Doak—Instructor, Agricultural Technology Program, Virginia Tech, Blacksburg, VA, United States.
Fred Fishel—University of Florida, Gainesville, FL, United States.
Carl Griffey—Department of Crop and Soil Environmental Sciences, Virginia Tech, Blacksburg, VA, United States.
Stephen Harrison—School of Plant, Environmental and Soil Sciences, LSU AgCenter, LA, United States.
Rocky Lemus—Mississippi State University, Mississippi State, MS, United States.
Nathan O'Berry—Pioneer Hi-Bred International, The Center for Dryland Research, Coastal Plains, Maize Development, Kinston, NC, United States.
Jeffrey S. Pippen—Nicholas School of the Environment, Duke University, Durham, NC, United States.
Peter M. Sforza—Director and Research Scientist, Geospatial Apps Development and Administration, Virginia Tech, Blacksburg, VA, United States.
Chris Teutsch—Southern Piedmont Agricultural Research and Extension Center, Blackstone, VA, United States.

About the Book

This book is designed to be unique in several ways. It provides in one place a description of the most common grasses, legumes, and nonleguminous forbs of the eastern United States. It covers many of the most important grassland, turf, and noncrop plants and their seeds. Included are both crops and weeds, since the two are inextricably connected in the field. Unlike many publications that include plant identification, we emphasize *Vegetative Identification*. Most plants flower for a relative short period; so the person in the field is frequently faced with identifying a plant without a flower. We also include *Floral Identification—because it can be definitive and can sometimes greatly simplify the identification process*. The photographs and other illustrations are intended to help with these determinations.

Besides identification of important species, we describe other key characteristics such as adaptation, favorable and unfavorable soil types, seasonal growth patterns, and toxicity. For plants harvested for hay or silage or by grazing, we discuss cutting and grazing management, quality factors, and potential yields.

Because of its organization and content, this book should be a valuable reference for farmers and farm advisors, teachers and students of agronomy, or for anyone interested in the dynamic relationship between plants and agriculture.

Legumes

Common Grasses, Legumes and Forbs of the Eastern United States. DOI: https://doi.org/10.1016/B978-0-12-813951-6.00001-7

Alfalfa—*Medicago sativa* L. subsp. *sativa*

Alfalfa seeds; kidney shaped

Seed

- Size: 2.0–2.5 mm long, 1.3–1.6 mm wide
- Shape and texture: typically kidney shaped; a notch on long side near the wide end, hilum located in the notch; surface smooth, dull, under high magnification appears textured
- Color: orange-brown or olive, area around hilum slightly lighter
- Seeds per pound: 220,000

Alfalfa plant

Alfalfa leaf toothed upper 1/3 (right) Sweet clover leaf toothed upper 2/3 or entire (left)

Leaf shape and arrangement—Leaves: Leaflets toothed on the upper 1/3 of the margins; leaflets of *sweet clover* toothed on the upper 2/3 of the margin or more; **Petiole**: petiolule of central leaflet distinctly longer than those of the two lateral leaflets; **Stipules**: with serrated margins and pointed tip.

Stolon/Rhizome/Roots: No stolons or rhizomes. Taprooted.

Inflorescence: Purple to bluish flower.

Alfalfa flower head *Source: John Wright*

Alfalfa—*Medicago sativa* L. subsp. *sativa*.

Synonyms: Lucerne
Family: Fabaceae/Leguminosae
Other common names: None reported
Life cycle: Cool-season perennial
Native to: Middle East (Iran)

Distribution and Adaptation
- Primarily found in Cool Humid and Transition regions, but grown to limited extent in Warm Humid region.
- Adapted to wide range of climatic conditions: can survive from below $-77°F$ with snow cover to above $120°F$ in California.
- Drought tolerant; will go dormant during severe drought but will survive.
- Not tolerant of poor drainage.
- pH >6.0, prefers 6.5–7.

Morphology/Growth Pattern
- Deep rooted up to 15 ft (4.6 m) in some soils, but most roots in transition region are in upper 3.0 ft (0.91 m).
- Can grow up to 3 ft (1 m).
- A mature alfalfa plant may have 5–25 stems.
- Fleshy part of root contains nonstructural carbohydrates needed for regrowth following harvest or grazing, dormancy and/or winter survival.
- New growth may occur from axillary stem buds as well as from crown buds.

Use and Potential Problems
- Bloat is a potential problem in cattle but not often in other domestic grazing animals.
- Not tolerant of frequent defoliation (less than 30 days is usually too frequent).
- Suggested management: first cut: mid bud to early flower.
  Regrowth harvests should be on 28–40 day intervals (minimum of 10% bloom).
- Final harvest may occur following killing frost, but leave 4–6″ stubble.
- Number of cuttings/year varies from 2 in northern area to 3–6 grazings or cuttings in the rest of region.
- Excellent quality: high protein, high minerals, and vitamin A.
- Annual yields range from 4 to 7 tons per acre/year throughout most of the area.

Toxicity/Disorders
- Bloating agents; cause bloat to cattle and sheep.

Similar Species
- At a vegetative stage, alfalfa and *sweet clover* (*Melilotus* spp) leaves resemble each other. However, alfalfa leaflets are toothed on the upper 1/3 of the margins while leaflets of *sweet clover* are toothed on the upper 2/3 or more of the margin. Alfalfa flowers are globe-like clusters unlike the flowers of *sweet clover* that are arranged in a raceme.

Reference
- Alfalfa [4,7,12–15,18,24–26]

Clover, Alsike—*Trifolium hybridum* L.

Alsike clover seeds; notch slightly off centered

Alsike clover stem and stipule

Alsike clover flower

Seed
- Size: 1.0—1.3 mm long, 0.9—1.3 mm wide 1/2 the size of alfalfa
- Shape and texture: nearly heart shaped except notch slightly off center; surface smooth but appears textured under high magnification
- Color: variable color: reddish brown, tan, greenish tan, may or may not have purple or black mottling
- Seed per pound: 680,000

Alsike clover plant

Leaf shape and arrangement—Leaves: Leaflets true trifoliate, finely toothed, both surfaces of leaflets glabrous (dull), oval tapering to a thin tip, with reticulate venation throughout; **Petiole**: short, diffused; **Stipule**: stipule at the base, large, lanceolate, pointed at the tip.

Stolon/Rhizome/Roots: No stolons or rhizomes; taproot with few branches.

Inflorescence: White to pink head; shorter peduncle than white clover; globe-like or nearly so seedhead.

Clover, alsike—*Trifolium hybridum* L.

Other common names: Swedish clover
Family: Fabaceae/Leguminosae
Life cycle: Cool-season perennial
Native to: Northern Europe

Distribution and Adaptation
- Grows best in a cool, moist climate with an abundance of soil moisture.
- Prefers heavy silt or clay soils, but will grow in all soils except sands.
- It tolerates acid and alkaline soils better than most clovers.
- Better suited to poor fertility soils than most clovers.
- Withstands cold weather better than red clover.
- A rotational grazing system where alsike is grazed to a height of 2−4 in. (5−10 cm) following a regrowth period of 4 weeks will result in a persistent stand.

Morphology/Growth Pattern
- Alsike clover has an upright growth habit and grows up to 1.25−2.5 ft (0.38−0.76 m) high; is a true trifoliate, no "V" blotch on leaflets like white and red clover leaflets; has taproot.
- Due to its fine stems it is susceptible to lodging during hay production.
- Average stand life of 2−3 years.

Use and Potential Problems
- Most useful in short-rotation pastures or in hay mixtures.
- Good quality forage.
- Report had indicated that when horses graze pastures with abundant alsike clover, they will have photosensitization problems. Photosensitization refers to reddening of the skin under the influence of sunlight, shortly followed by shallow or deep dry skin necrosis. Symptoms can also include digestive and nervous system disorders.
- Can cause bloat.

Toxicity/Disorders
- Photosensitization agents.

Similar Species
- Alsike clover resembles both red clover (*Trifolium pratense* L.) and white clover (*Trifolium repens* L.) in appearance. However, alsike clover differs from red clover in that it is glabrous, its leaves do not have white markings and the flowers of alsike clover are pink to white while *red clover* flowers are red to purple and have two short stocked leaves directly below the flowerhead. White clover unlike alsike clover is stolonferous, leaves are shiny underside, long peduncles, and petioles almost equal size.

Reference
- Alsike clover [4,7,12,15,19,26]

Clover, Arrowleaf—*Trifolium vesiculosum* Savi

Arrowleaf clover seeds

Arrowleaf clover (left) and red clover (right)

Arrowleaf clover flower head

Seed
- Size: 1.0–1.5 mm long, 0.7–0.9 mm wide, seeds are about half the size of alfalfa seed and twice the size of white clover seed
- Shape and texture: oval-shaped; fine textured
- Color: yellow to dark brown
- Seeds per pound: 400,000

Arrowleaf clover plant

Leaf shape and arrangement—Leaves: Leaflets true trifoliate, very pointed, smooth (no hair); leaves have very pronounced veins; leaflet normally arrow-shaped with a large white "V" shaped mark on many leaves; both arrowleaf clover and red clover have water marks on the leaf, red clover leaf however is more rounded and hairy; **Petiole**: long petiole; **Stipules**: long, white pointed stipules at the base of each leaf petiole.

Stolon/Rhizome/Roots: No stolons or rhizomes; stem hollow, thick, often purple; fibrous root systems.

Inflorescence: Flowers white initially, later change to pink to purple; flower heads on long peduncles; flower heads are conical and several inches long.

Clover, arrowleaf—*Trifolium vesiculosum* Savi

Other common names: None recorded
Family: Fabaceae/Leguminosae
Life cycle: Winter annual
Native to: Mediterranean region

Distribution and Adaptation
- Commonly found in the warm humid zone but limited use in the southern edge of the transition zone.
- Best suited to well-drained, sandy soils, with pH 6—7.
- Cold and drought tolerant.

Morphology/Growth Pattern
- Forms a leafy rosette during early growth and then produces branching stems that curve upward to lengths of 1.67—2.33 ft (51—71 cm).
- As plant matures it may become "lodged" and intertwined, if not harvested.
- Reseeding potential is good due to the high percentage of hard seeds (90%) that are produced.
- If managed well, continues to develop new leaves and remains fairly leafy and productive longer in the spring.

Use and Potential Problems
- Grown primarily as a grazing crop for domestic livestock, but also as a cover crop.
- Browsed by white-tailed deer, turkey and other wildlife.
- Virus diseases limit its and poor stand establishment.

Toxicity/Disorders
- Bloating agents; cause bloat to cattle and sheep.

Similar Species:
- Both *red clover* (*Trifolium pratense* L.) and *arrowleaf clover* have "V" water marks on the leaf, red clover leaf however is more rounded and hairy while arrowleaf clover leaves are arrow-shaped and are glabrous.

Reference
- Arrowleaf clover [5,7,12,29—31,63,64]

Clover, Crimson—*Trifolium incarnatum* L.

Crimson clover seeds

Seed

- Size: 1.9–2.3 mm long, 1.4–1.6 mm wide
- Shape and texture: oblong, football shaped, larger than most clover seeds and larger than alfalfa; smooth textured, glossy, minutely pitted
- Color: cream to yellow can also be red
- Seeds per pound: 140,000

Crimson clover plant (Crimson clover (upper), red clover (lower)) *Source: Chris Teutsch*

Crimson clover leaf *Source: Chris Teutsch*

Leaf shape and arrangement—Leaves: Leaflets true trifoliate, are usually unmarked but sometimes have a few dark-red spots; dark green leaves densely covered with hairs, leaflets slightly toothed, softly pubescent, resembles red clover, but its leaves have a more rounded tips; **Petiole:** petiole of some lower and medium leaves longer, petiole of upper leaves shorter and pubescent; **Stipule:** shallowly toothed and purplish tipped.

Stolon/Rhizome/Roots: No stolons or rhizomes; taproot and fibrous branch roots.

Inflorescence: Flowers are crimson, roughly 15.9 mm long by 3.2 mm wide, flower opens in succession from the bottom to the top of the flower head.

Crimson clover flowerhead *Source: Chris Teutsch*

Clover, crimson—*Trifolium incarnatum* L.

Other common names: Scarlet clover, Italian clover, Incarnate clover
Family: Fabaceae/Leguminosae
Life cycle: Winter annual
Native to: Atlantic and southern Europe

Distribution and Adaptation
- Grown primarily in the warm humid zone but also in the transition zone.
- Will grow on most soil types that are well drained, ranging from sands to heavy clays. Poorly drained, muck or water logged soils not suitable.
- Best growth occurs on loams with a good organic matter content.
- Will grow on soils varying in acidity and alkalinity but not on extremely acid soils.
- Can grow on soils with pH of 5−8; optimum growth and nodulation occur between pH of 5.5 and 7.5.
- Does not survive extreme heat or cold.
- Prefers cool, humid weather and requires 35 in. (889 mm) or more of rainfall/year.
- Prefers full sun but can tolerate partial shade.

Morphology/Growth Pattern
- Reaches a height of 1−2 ft (0.30−0.60 m).
- The root system consists of a central taproot and fibrous branch roots.
- Volunteer reseeding is important and varieties vary in the amount of hard seeds produced and they are classified as hard or soft seed producers. Soft-seeded varieties potentially allow for germination during summer which may suffer death loss before the favorable autumn season.
- Has excellent seedling vigor and is the best of the winter annual clovers for early forage production.

Use and Potential Problems
- Used for livestock feeding and browsed by white-tailed deer, turkey, and other wildlife.
- The plant is widely used as forage, pasture, green manure, cover crop, vineyard cover, and erosion control.
- Due to its earliness, it is excellent in no-till rotations with corn and sorghum.
- Grazing may begin earlier than for most annual clovers because of high fall and winter growth.
- Because of its good seedling vigor and early spring maturity, it is ideal for overseeding warm-season perennial grasses such as bermudagrass and Bahiagrass for winter and spring grazing.
- Also grown with cereals (rye, wheat, barley, oats, triticale, and annual ryegrass).
- Good quality hay can be obtained at 50% or less bloom. Lack of good drying weather in the early season can make this practice difficult.
- The advantages of crimson clover over hairy vetch is that it grows quicker, is a better weed suppressor in the fall, and is earlier to mature in the spring.
- Hairy vetch is better adapted than crimson clover to deep sandy soils.
- Disadvantages of crimson clover include short growing season, somewhat unpalatable when grazed and low quality when mature.

Toxicity/Disorders
- Bloating agents; causes bloat to cattle and sheep.

Reference
- Crimson clover [5,7,12,22,29−31,57,63,64]

*Clover, Large Hop—*Trifolium campestre* Schreb.; Clover, Small Hop—*Trifolium dubium* Sibth.

Large hope clover seeds

Large hop clover leaves, stem, peduncles, and flowers; peduncles often exceeding the subtending leaves

Seed
- Size: $1.5 \times 1.2 \times 1$ mm; 1/3 size of alfalfa, like miniature crimson clover
- Shape and texture: oval; flattened, sides slightly curved; very glossy
- Color: yellow and red colored
- Seeds per pound: 900,000−1,000,000

Large hop clover plant

Leaf shape and arrangement—Leaves: (Large hop clover): Leaflets not true trifoliate, toothed from mid-blade to tip; the leaves of large hop clover are similar to small hop clover but have larger leaves and flower heads than small hop clover (*Trifolium dubium* Sibth.). **Petiole**: the terminal leaflet of large hop clover is stalked (petioled); **Stipule**: large, thin, ovate; large hop clover peduncles are stout, stiff, and straight, often longer than subtending leaves.

Stolon/Rhizome/Roots: No stolons or rhizomes; upright, hairy branched stems.

Inflorescence: Large hop clover has bright yellow flowers 20−30 in loose clusters on long stalks attached at leaf axils; flowers turn brown and fall-back when mature.

---

*Plant and seed description for large hop clover unless indicated otherwise.

Clover, large hop—*Trifolium campestre* Schreb.

Other common name: Low hop clover
Family: Fabaceae/Leguminosae

Clover, small hop clover—*T. dubium* Sibth.

Other common name: Least hop clover
Life cycle: Winter annual
Native to: Europe

Distribution and Adaptation
- Frequently found in the whole region, but most often in the warm humid and transition zones.
- Found on roadsides, waste places, and pastures with low soil fertility.
- Adapted to droughty, acid, and eroded soils.

Morphology/Growth Pattern
- Erect with ascending branches; growth to height of 0.2−1.0 ft (0.05−0.30 m).
- Leaflets of hop clover are unlike most other clovers in that the center leaflet has a longer petiolule than the other two leaflets, like species in the *Medicago* and *Melilotu*s genera; seedheads are smaller than other clovers; globe-like inflorescence; petals yellow, turning brownish when dry; flower heads have 20−30 (fewer in *T. dubium* L.) yellow-ish flowers that turn light brown at maturity.

Use and Potential Problems
- Both species are used primarily in less productive pastures and to a lesser extent in hay fields.
- They are not planted but volunteer from year to year.
- Both species have high nutritive values and are palatable.
- Both species reproduce through volunteer reseeding in pastures especially in absence of other legumes.
- They are relatively low yielding over a short growing season.
- Need to be utilized quickly due to a short season of production.
- Powdery mildew is a common disease of these species.

Toxicity/Disorders
- Bloating agents; causes bloat to cattle and sheep.

Similar Species
- The florets of *small hop clover* (*T. dubium* L.) are shaped like a "baseball bat" and arranged in loose clusters; while *large hop clover* (*T. campestre* L.) individual flowers are more rounded in shape and flare at the tip.

Reference
- Large hop clover [5,12,17,29−31,44,47,57,72]

Clover, Rabbitfoot—*Trifolium arvense* L.

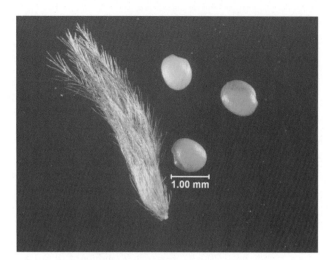

Rabbit foot clover seeds

Seed
- Size: 0.9–1.3 × 0.6–0.7 × 0.5 mm
- Shape and texture: mitten shaped, compressed; semi-glossy
- Color: yellow to tan

Rabbitfoot clover plant

Rabbitfoot clover hairy leaves and stems

Leaf shape and arrangement—Leaves: Leaflets true trifoliate, leaves are composed of three long, narrow leaflets and they are attached to the spreading branches by short stems; both stem and leaves densely hairy; **Petiole**: long, pointed paper; **Stipule**: thin stipules cover the junction between the leaf and stem.

Stolon/Rhizome/Roots: No stolons or rhizomes; taproot with secondary fibrous root systems.

Inflorescence: Flowers are pinkish-white sometime grayish and form short, cylindrical heads; head resembles rabbit's foot.

Rabbitfoot clover flower head

Clover, rabbitfoot—*Trifolium arvense* L.

Other common names: Stone clover, old-field clover, hare's foot clover
Family: Fabaceae/Leguminosae
Life cycle: Winter annual
Native to: Europe and North Africa

Distribution and Adaptation
- Found in warm humid and southern part of the transition zones.
- Habitat: roadsides, low maintenance areas, disturbed areas, and occasionally in low performing pastures.
- Adapted to infertile, dry sandy soils where other clovers are not found.
- Often found in association with low growing numerous cool season annual grasses.

Morphology/Growth Pattern
- Has a multibranched growth habit; it is erect.
- Grows from 4 to 16 in. (10−41 cm) in height.
- Reproduces by seed.

Use and Potential Problems
- The lack of competitiveness with other species is one of the reasons why this clover is not productive.
- If grown under favorable conditions, it may produce 2000 lbs/acre in late spring.
- Annual production potential is low.
- In most situations, grazing the growth is more practical than harvesting because of low yields.

Toxicity/Disorders
- Bloating agents; cause bloat to cattle and sheep.

Reference
- Rabbitfoot clover [5,7,12,29−31,44,47,63,64,72]

Clover, Red—*Trifolium pratense* L.

Red clover seeds; color often blends from dark to light within one seed

Seed
- Size: 1.9–2.3 mm long, 1.4–2.0 mm wide
- Shape and texture: boxing glove shaped; thumb is 1/4 as long as mitt, usually fatter than sweet clover; surface smooth, slightly glossy or dull
- Color: straw colored, with or without purple mottles, to solid light purple, often blends from dark to light within one seed
- Seeds per pound: 275,000

Red clover plant *Source: Chris Teutsch*

Red clover hairy stem, stipule and petiole

Leaf shape and arrangement—Leaves: Leaflets true trifoliate, leaves entirely pubescent on both sides, leaflets usually have light colored crescentic "V" shape marking on leaves; leaf margins usually ciliate; **Petiole**: leaves of the basal rosette have long petiole (stalked), while those leaves on the stem have relatively long petioled to nearly sessile; **Stipules**: large hairy stipules connected to petioles for half their length.

Stolon/Rhizome/Roots: No stolons or rhizomes; taproots the first year, the second year plants develop secondary roots.

Inflorescence: Two short stocked leaves directly below flower head, sessile or very short pedunculate.

Red clover (lower) and crimson clover (upper) flowerheads

Clover, red—*Trifolium pratense* L.

Family: Fabaceae/Leguminosae
Types: Early flowering (medium red often referred to as double-cut), and late flowering (American mammoth often referred to as single-cut)
Life cycle: short lived cool-season perennial or biennial
Native to: Southeastern Europe and Turkey

Distribution and Adaptation
- Primarily grown in the cool humid and transition zones, but is often grown in the upper parts of the warm humid zone.
- Adapted to cool summers with adequate moisture.
- Moderately tolerant of poor drainage, but not as tolerant as white clover but better than alfalfa.
- Is shade tolerant.
- Will tolerate acid soil better than alfalfa, for best result 6.0−7.0.

Morphology/Growth Pattern
- Erect; growth to the height of 1.3 ft (40 cm).
- The late maturing type does not produce flowering stems the first year, instead produces a rosette type growth the first year. The second year, it blooms 10−14 days later than the early flowering (medium red) type.
- Herbaceous plant with numerous leafy stems growing from the crown.
- Less apical dominance in red clover than in alfalfa.
- Usually favored when rest periods following defoliations are 14−30 days or the canopy reaches 12−16 in. (30−41 cm) in height. Low to medium tolerance to frequent grazing/cutting.
- Vigorous seedling growth.

Use and Potential Problems
- Used primarily in mixture with grasses for hay and grazing.
- It is most often used in mixtures with cool season grasses, but may be grown as a pure stand, especially for hay. The early flowering type (medium red clover) produces two or three hay crops per year. The late flowering type (American mammoth red clover) or single-cut type usually produces one crop.
- It is excellent quality when grazed or harvested in preflower stage of growth.
- Does cause bloat, especially in cattle.
- Black patch fungal disease caused by *Rhizoctonia leguminicola* actual alkaloid. The disease is the metabolite of the *R. leguminicola*. Cattle and horses grazing or feeding hay clovers or alfalfa infected by *R. leguminicola* will show excessive salivation, slobbering, and weight loss; recovery can occur rapidly once animals are moved from the infected field. The compound in infected hay will decline in time. Excessive moisture and high humidity will increase the occurrence of black patch fungal disease, more prevalent when high rainfall and humidity is present.

Toxicity/Disorders
- Alkaloid from black patch disease on *Trifolium pratense*. Bloating agents.

Reference
- Red clover [5,7,12,17,22,29−31,57,63,64,72]

Clover, Subterranean—*Trifolium subterraneum* L.

Subterranean clover seeds

Hairy stem and leaf of subterranean clover

Subterranean clover flower

Seed

- Size: 2.3–3.2 × 2.2–2.6 × 2 mm
- Shape and texture: mitten shaped, smooth textured
- Color: dark reddish brown, some nearly black or white
- Seed per pound: 65,000

Subterranean clover plant

Leaf shape and arrangement—Leaves: Leaflets true trifoliate, leaflets mostly heart-shaped, finely toothed with the notch at the tips of the leaflet; hairy over most of its surface leaves; long-petioled; **Petiole**: leaves long-petioled; **Stipules**: stipule roughly tapered, veined, and often has a reddish coloring diffused across its surface.

Stolon/Rhizome/Roots: No stolons or rhizomes; stem hairy, prostrate, and not rooting.

Inflorescence: White flowers, but become pink as it approaches maturity; three or four fertile flowers grouped together with some infertile flowers at the end of a stalk rising from the axil of the leaf.

Clover, subterranean—*Trifolium subterraneum* L.

Other common name: Subclover
Family: Fabaceae/Leguminosae
Life cycle: Winter annual
Native to: Mediterranean and southern England

Distribution and Adaptation
- Best adapted to the warm humid region and southern parts of the transition zone.
- Best adapted to fine sandy loams to clay.
- Tolerates moderate acid soil conditions and somewhat poorly drained soils.
- Has low shade tolerance and reseeds well in many environments.

Morphology/Growth Pattern
- Grows 6−10 in. (15−25 cm) tall.
- After flower pollination, the flowers "peg down" and the seeds develop on or just beneath the soil surface.
- Subterranean clover has a low growth habit and forms a dense sod.
- Reproduces by seed.

Use and Potential Problems
- Used almost solely as a pasture legume in mixtures with grasses, especially with warm season grasses during their dormant period.
- It has good reseeding characteristics and has been used as a living mulch.
- Like arrowleaf clover, it becomes chlorotic and stunted on soils with a pH above 7.3.
- Its estrogenic content can potentially cause problems in reproductive animals.

Toxicity/Disorders
- Phytoestrogens.

Reference
- Subterranean clover [5,7,12,22,29,30,31,63,64]

Clover, Yellow Sweet—*Melilotus officinalis* (L.) Lam.; Clover, White Sweet—*Melilotus alba* Medik.

Yellow sweet clover seeds

Leaflets serrated over entire margin

Yellow sweet clover flower

Seed
- Size: 1.7—1.9 mm long, 1.0—1.3 mm wide
- Shape and texture: ovate, baseball mitt shaped, thumb is 2/3 as long as mitt, parallel crease; surface smooth, dull
- Color: yellow-brown, greenish or orangish, above the hilum small dark spot
- Seeds per pound: 260,000

Sweet clover plant *Source: Lachlan Cranswick*

Leaf shape and arrangement—Leaves: Leaflets not true trifoliolate, leaflets serrated over entire margin, central leaflet on long petiolule and the two lateral leaflets on a short or no petiolule; **Petiole**: all leaflets on short petiolules; **Stipules**: small, papery, smooth (not toothed) 7—10 mm long.

Stolon/Rhizome/Rroots: No stolons or rhizomes; often much branched taproot and erect or ascending stems, glabrous or slightly pubescent.

Inflorescence: Flowers arranged on long racemes; flowers on recurved pedicles; sweet odor; white or yellow flowers; petals pure white; raceme is long compared to alfalfa; yellow sweet clover is similar in appearance to white sweet clover (*Melilotus alba*) except the white sweet clover has white instead of yellow flowers.

Clover, yellow sweet—*Melilotus officinalis* (L.) Lam.: **white sweet clover**—*M. alba* Medik.

Other common names: King's crown, plaster clover, Hart's clover, king's clover
Family: Fabaceae/Leguminosae
Life cycle: Cool-season biennial
Native to: Mediterranean region

Distribution and Adaptation
- Found primarily in the cool humid zone and to lesser extent in the transition zone.
- Grows under a wide range of soil and climatic conditions.
- Best adapted to alkaline or calcareous soils; does not tolerate acid soils.
- Is drought resistant.
- Found in disturbed places (construction, mining, and roadsides) where nonacid subsoil is exposed.

Morphology/Growth Pattern
- Yellow sweet-clover is biennial with several stems ascending from 2 to 10 ft (0.50−3.0 m) high.
- In the seeding year, the growth habit of sweet clover is different from alfalfa and many clovers; in the seedling year sweet clover produces a single, well-branched stem and near the end of the first growing season, several buds form at the crown. Near the end of the first growing season, the root enlarges and serves as a storage organ for carbohydrate reserves.
- Spring of the second year, new growth arises from the crown buds much the same as alfalfa.
- Seeds of sweet clover are hard, and need scarification before they will germinate.

Use and Potential Problems
- Is used for green manure crop, hay, pasture, and silage.
- Is excellent source of nectar for honeybees.
- A very important soil improving crop.
- Good forage source early, but as the plants mature they decline rapidly in feed value as they accumulate coumarin.
- Coumarin is a bitter compound that gives the plant its characteristic sweet aroma, especially upon drying: this reduces its palatability to animals.
- Coumarin may be converted to dicoumarol (anticoagulant) in moldy hay or drought stressed plants and may result in hemorrhaging in animals consuming it.
- Varieties with low coumarin content have been developed.
- May also cause bloat, especially in cattle.

Toxicity/Disorders
- Coumarin, which is converted to hydroxycoumarin and dicoumarol by fungi in partially cured stems.

Similar Species
- At a vegetative stage, *sweet clover* and alfalfa (*Medicago sativa* L.) leaves resemble each other, however, alfalfa leaflets are toothed on the upper 1/3 of the margins while leaflets of *sweet clover* are toothed on the upper 2/3 of the margin or more. Alfalfa flowers are global unlike the flowers of sweet clover that are arranged in a raceme.

Reference
- Yellow sweet clover [5,7,12,17,29−31,39,44,47,61,63,64]

Clovers, White (Small, Intermediate, and Large (Ladino) Types)—*Trifolium repens* L.

White clover seeds

White clover stolons *Source: NCSU*

White clover flower *Source: Lachlan Cranswick*

Seed

- Size: 0.9–1.1 mm long, 0.8–1.1 mm wide
- Shape and texture: rounded and somewhat heart shaped; about the same size or smaller than alsike clover; surface smooth, dull; under high magnification there is a slight sheen
- Color: usually entirely yellowish, sometimes reddish
- Seeds per pound: 800,000

Kudzu plant *Source: Chris Teutsch*

Leaf shape and arrangement—Leaves: Leaflets true trifoliate; leaf palmately trifoliate, usually with white or yellow mark with "M" or "W" and sometimes with dark red flecks, leaflets glabrous with very long petioles; leaflet heart-shaped with white crescent mark; the undersides of white clover are frequently shiny distinguished by the lateral veins of the leaflets straight toward the margin, wide spaced; **Petiole**: very long petiole, grooves on upper surfaces of petioles; **Stipule**: closed stipules forming a membranous sheath.

Stolon/Rhizome/Roots: Stolons; no rhizomes; stem glabrous (hairless); taprooted.

Inflorescence: White flowers on long peduncles.

Clovers, white (small, intermediate, and large (Ladino) types)—*Trifolium repens* L.

Types: Small, Intermediate, and Large.
Family: Fabaceae/Leguminosae
Life cycle: Cool-season perennial
Native to: Eastern Mediterranean region

Distribution and Adaptation
- Most widely adapted legume in the United States. It is found in the entire region, with the possible exception of south Florida.
- Common in pastures, lawns, fields, roadsides, dry meadows, and wood margins.
- Does better on clay or loam soils than on sandy soils and grows best on well-drained loam or clay soils.
- Small, intermediate, and large white clovers are adapted to infertile, medium fertile, and fertile soil, respectively.
- Grows best under cool, moist conditions and is winter hardy.
- Performance is best when soil pH is in the range of 6.0−7.0 and with high phosphorus and potassium availability.
- Not tolerant of saline or highly alkaline soils.

Morphology/Growth Pattern
- Plant height depends on type.
- White clovers are classified primarily by size (small, intermediate, and large). Many forms/types, small white clover (*T. repens* L. f. *repens*), intermediate white clovers (*T. repens* f. *hollandicum* Erith ex Jav. & Soo) larger than white Dutch. And large white clover, or ladino clover (*T. repens* var. *giganteum* L.).
- The main differences between white clover types are differences in height, growth, size of leaves, flower heads, stolons, and length of internode.
- Has a shallow root system (over 80% located in the top 7−8 in. (18−20 cm) of the soil surface), with stolons rooting at the nodes where they contact the soil thus often forming a mat.
- Can develop a taproot, which often dies in 1 or 2 years leaving the secondary roots to develop from the stolons that become the main root system. Stolons are dependent on its adventitious root systems.
- Plants regenerate themselves by reseeding and by stolons.
- The intermediate types can reseed readily while the ladino clovers usually are not as dependable to reseed.

Use and Potential Problems
- The Dutch white clover and other improved intermediate cultivars can tolerate close grazing. However, under close continuous stocking, leaf size is often reduced and more prostrate growth can occur due to more stolon branching.
- The intermediate and large white clovers are most often used as a grazing plant in mixture with grasses. The large (ladino) types on occasion might also be successfully cut for hay about once a year in late spring.
- The large white clovers respond well to rotational stocking.
- Manage to maintain 50%−60% grass and 40%−50% clover.
- White clover has a high potential for causing bloat.
- Is a high-quality forage legume partly due to the fact that animals graze mostly leaves, petiole, peduncle, and flowers making it the most nutritious forage available.
- Has higher crude protein than birdsfoot trefoil, alfalfa, or red clover. It is also very palatable and highly digestible. Research showed crude protein content of white clover in excess of 30% and the digestibility 80%.
- Used for livestock grazing, soil improvement, and erosion control.
- Dry matter yields depend on the type of white clover ranging from 2000 pounds per acre per season for white Dutch clover to 6000 for Ladino clover in mixture with grasses.

Toxicity/Disorders
- Bloating agents; Phytoestogens.

Similar Species
- Small *white clover* has small leaves, leaflets, short petioles, short peduncles, and small flowers; the intermediate type is intermediate in size; the large white clover (Ladino clover) has the largest leaves, leaflets, petioles, and peduncles.

Reference
- White clovers [5,7,12,13,17,22,29 31,39,44,45,57,59,63,64,67,72]

Clover Summary

Leaves and Flowers

Alsike clover

Alsike clover

Arrowleaf clover

Arrowleaf clover

Crimson clover

Crimson clover

Crimson clover

White clover

White clover

Red clover

Red clover

Red clover

Clover Summary

Leaves and Flowers

Rabbitfoot clover

Rabbitfoot clover

Subterranean clover

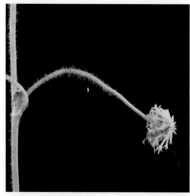

Subterranean clover

Sweet clover

Sweet clover

Kudzu—*Pueraria lobata* (Willd.) Ohwi var. *lobata.* = *Pueraria montana* (Lour.) Merr. var. *lobata* (Willd.) Maesen & S.M. Almeida

Kudzu seeds

Kudzu leaves *Source: Peter Sforza*

Kudzu flowers *Source: Peter Sforza*

Seed
- Size: 2.8−4 × 2−2.7 × 2 mm
- Shape and texture: compressed or nearly so, glossy, smooth with oblong, marginal hilum with rim
- Color: reddish brown mottled with black
- Seeds per pound: 40,000−45,000

37,000 seeds per pound

Kudzu on power line *Source: Chris Teutsch*

Leaf shape and arrangement—Leaves: All leaflets occur on petioles, however the lateral leaflets are on very short **petioles** (less than 1.3 cm), while the center leaflet occurs on a **petiole** approximately 2 cm long; short **stipule**; large trifoliolate leaves; leaflets 5−8 cm in diameter arranged alternately along the stem; center leaf somewhat diamond-shaped, all leaflets are lobed, however, the lateral leaflets are usually lobed on one side only, while the center leaflet is usually lobed on both sides.

Stolon/Rhizome/Roots: Stoloniferous, no rhizomes, deep-rooted; stems covered with brown hairs, when young, stems are covered with stiff bronze hairs but not common on older plants.

Inflorescence: Flowers in racemes that arise from the axillary regions, each raceme is approximately 10−20 cm long and is reddish purple in color, flowers are often inconspicuous due to the dense vegetation that often covers them.

Kudzu—*Pueraria lobata* (Willd.) Ohwi var. *lobata.* = *Pueraria montana* (Lour.) Merr. var. *lobata* (Willd.) Maesen & S.M. Almeida

Other common names: Mile-a-minute vine, foot-a-night vine, and the vine that ate the South
Family: Fabaceae/Leguminosae
Life cycle: Warm-season perennial
Native to: Japan and China

Distribution and Adaptation
- Found mostly in the warm humid and lower parts of the transition zones.
- Introduced to the United States as a soil cover to prevent erosion in 1876.
- Grows on a wide range of soil types.
- Tolerant of low fertility and soil acidity but not poor drainage.
- Is drought tolerant
- Not really adapted to shade nor frequent defoliation.

Morphology/Growth Pattern
- Climbing or trailing, perennial vine.
- Vine can grow as much as a foot per day during summer months.
- Propagation by seeds but primarily by stolons.
- The leaves maintain good quality even if not grazed for long periods of time.

Use and Potential Problems
- May be used as pasture or harvested for hay.
- High nutritive values and palatable to livestock
- Kudzu is often a weed in forests and of rights-of-way for power lines and roadsides.

Toxicity/Disorders
- Potential bloating agent.

Reference
- Kudzu [7,12,22,44]

Lespedeza, Kobecommon—*Kummerowia striata* (Thunb.) Schindl. "Kobe," Also Called *Lespedeza striata* (Thunb.) Hook & Arn.

Kobe lespedeza seeds

Seed
- Size: 1.7—2.3 mm long, 1.2—1.7 mm wide
- Shape and texture: oblong, compressed; can see network of veins in the seed pod
- Color: black and shiny without hull when in pod, appears gray to papery brown
- Seeds per pound:185,000

From left to right Korean lespedeza and Kobe lespedeza plants

Kobe lespedeza leaf

Leaf shape and arrangement—Leaves: Common and Kobe leaves are longer and less rounded on the ends than Korean lespedeza. Leaves of the common type do not fold around the seed pod, so they are more susceptible to shattering; **Stipules**: much smaller than on Korean lespedeza.

Stolon/Rhizome/Roots: No stolons or rhizomes; fine stemmed; hairs on stems downward on Kobe lespedeza while hairs on stems upward on Korean lespedeza; shallow taproot systems.

Inflorescence: Vary in color from light blue to purple; the seeds of common lespedeza are set in leaf axils, directly along the entire length of the stem, while the seeds of Korean lespedeza grow in clusters at the terminal end of each branch.

From left to right Korean lespedeza and Kobe lespedeza leaves
Source: Peter Sforza

Lespedeza, kobe—*Kummerowia striata* (Thunb.) Schindl. "Kobe," also called *Lespedeza striata* (Thunb.) Hook & Arn.

Other common names: Striate lespedeza, common lespedeza, Japanese lespedeza, Japanese clover
Family: Fabaceae/Leguminosae
Life cycle: Warm season annual
Native to: Japan

Distribution and Adaptation
- Primarily in the warm humid and transition zones and to a limited extent in the extreme lower region of cool humid.
- The limits of Kobe lespedezas adaptation westward is limited by moisture, while the extent of the northern boundaries is limited by the plants ability to produce seeds before a killing frost.
- Tolerant of lower fertility and acid soils, however, responds well to lime and phosphorous fertilization.
- Is drought resistant.

Morphology/Growth Pattern
- Kobe lespedeza is an erect plant that grows 4.0−4.5 ft (1.22−1.37 m) in height.
- Two types of striate lespedezas, common and Kobe lespedezas; Kobe is the most widely used and most familiar cultivar of common lespedeza. Kobe is larger and taller than the common striate.
- Depending on where it is grown, both common and Kobe lespedezas (*Striate lespedezas*) reseed themselves. Korea lespedeza is mostly used in the northern portion of the lespedeza region, while straite lespedeza (common/Kobe) is commonly adapted toward the southern lespedeza region.

Use and Potential Problems
- Excellent for both hay and pasture for all domestic livestock.
- Makes excellent hay if the leaves can be retained during the hay making process; they tend to shatter very easily when the forage is dry enough to bale.
- Excellent food for deer, turkey, cottontail rabbit, quail, and dove.
- Used as a companion legume in tree plantings.
- Also used to stabilize soils on side slopes, or any roads crossing streams or drainages to prevent erosion or sedimentation; makes an excellent ground cover which controls erosion.
- Does not cause bloat.

Toxicity/Disorders
- None recorded.

Similar Species
- Both striate (common and Kobe) and *Korean lespedeza* (Kummerowia *stipulacea* are fine stemmed and shallow rooted. The leaves of *Korean lespedeza* are broad, heart-shaped and are distinctly veined, while common/Kobe lespedeza leaves are narrower, longer and less rounded on the ends. Also, *Korean lespedeza* stipules are much longer than *common/Kobe lespedeza*. After flowering, leaves of *Korean lespedeza* turn forward around the seed pod, which keeps the seed from shattering, unlike *Korean lespedeza*, the leaves of common/Kobe lespedeza do not fold around the seed pod thus making it more susceptible to shattering.

Reference
- Kobe lespedeza [5,7,17,22,29−31,63,64]

Lespedeza, Korean—*Kummerowia stipulacea*, (Maxim.) Makino, Also Called *Lespedeza stipulacea* Maxim.

Korean lespedeza seeds

Stipules large, paper thin; leaves distinctly veined *Source: Nathan O'Berry*

Korean lespedeza leaves, stem and flower *Source: Nathan O'Berry*

Seed

- Size: 1.7–2.2 mm long, 1.4–1.7 mm wide
- Shape and texture: oblong, compressed; surface smooth, glossy, may see grooves on surface, network of veins in the seed pod
- Color: dark brown and shiny without hull; pod, appears gray to papery brown
- Seeds per acre: 240,000

Sericea lespedeza (left), Korean lespedeza (right) plant

Leaf shape and arrangement—Leaves: Trifoliolate leaves; leaflets broad and heart-shaped are distinctly veined, while common and Kobe leaves are longer and less rounded on the ends, leaflets may have brownish spots on them; noticeable parallel venation on leaflets branching from midvein; **Petiole**: leaflet on a very short petiole; **Stipules**: large paper-thin light green stipules, oviate.

Stolon/Rhizome/Roots: No stolons or rhizomes; stems short pubescent; hairs on stems point up.

Inflorescence: Purplish-blue flowers born in short, dense spike-like racemes, usually arising from the axils of the upper leaves and a few from the lower nodes; seedhead hidden by the bract-like stipules.

Lespedeza, Korean—*Kummerowia stipulacea*, (Maxim.) Makino, also called *Lespedeza stipulacea* Maxim.

Other common names: Korean clover, Korean bush-clover
Family: Fabaceae/Leguminosae
Life cycle: Warm season annual
Native to: Korea (introduced into the United States in 1919).

Distribution and Adaptation
- Primarily in the transition zone and to lesser extent in the northern part of warm humid zone and in the extreme lower region of cool humid.
- Grows on acidic, poor fertility, and mined soils.
- Responds well to fertile, limed, and well-drained soils.
- Korean lespedeza is less tolerant of acidic soils and more tolerant of alkaline soils compared with striate lespedeza (common or Kobe).

Morphology/Growth Pattern
- Generally, Korean cultivars mature earlier than the striate cultivars (common and Kobe).
- Both Korean and striata species are fine stemmed, leafy with shallow taproot.
- Grows to a height of 2–3 ft (61–91 cm).
- Mostly self-pollinated.
- It is a good reseeding annual.

Use and Potential Problems
- Excellent for both hay and pasture for all domestic livestock.
- Excellent forage quality when grazed or when leaves can be retained during the hay making process.
- Does not cause bloat.

Toxicity/Disorders
- None recorded

Reference
- Korean lespedeza [7,17,22,29–31,63,64]

Lespedeza, Sericea—*Lespedeza cuneata* {Dum. Cours.} G. Don

Sericea lespedeza seeds

Seed

- Size: 1.6—1.8 mm long, 1.0—1.3 mm wide
- Shape: ovate with a shallow notch near the wide end, compressed; surface smooth, somewhat glossy, may be a faint groove near the hilum
- Color: yellowish tan to reddish brown, mottled or not with purple spots
- Seeds per pound: 335,000

Sericea lespedeza (left), Korean lespedeza (right)

Sharp point end of leaflets

Leaf shape and arrangement—Leaves: Trifoliate leaves distributed along the entire stem, leaflets are much longer than wide, linear, and narrow; the tip of the leaflets are slightly indented and end in a sharp point; leaflets have many short silky hairs on both surfaces and have a grayish cast; **Petiole**: the lower leaves have petioles, but the upper leaves are nearly sessile; **Stipules**: has scale-like stipules on stems.

Stolon/Rhizome/Roots: No stolons or rhizomes; coarse to fine stem, somewhat woody with age and have stiff bristles; tap rooted.

Inflorescence: Flower head emerge from leaf axils in the middle to upper portions of the plant; flowers appear single or in clusters of 2—4; flower colors vary from cream to purple.

Sericea lespedeza stem, leaves and flowers

Sericea lespedeza—*Lespedeza cuneata* (Dum. Cours.) G. Don

Other common name: Chinese bush clover, Chinese lespedeza
Family: Fabaceae/Leguminosae
Life cycle: Warm-season perennial
Native to: Eastern China, Korea and Japan

Distribution and Adaptation
* Grown primarily in the warm humid and transition zones from eastern Texas, Oklahoma, and Kansas eastward through Missouri, southern Illinois, and Indiana to the Atlantic Coast.
* Drought tolerant.
* It does best on clay and loamy soils that are deep, fertile, and well drained, but will also grow on shallow and infertile soils.
* Tolerates acid soils with aluminum levels toxic to many species.
* Grows well on low fertility soils where other legumes will not survive, such as mine spoils and rights of way with pH of 5 or below.
* Is best adapted to a pH of 6.0−6.5.

Morphology/Growth Pattern
* Has an erect growth habit, may reach 3−5 ft (91−152 cm) in height depending on soil types.
* Stems may be single or clustered and contain numerous branches.
* Deep branched tap root.
* Plant top growth is killed by frosts, and new growth is initiated in spring from crown buds.
* Has weak seedling vigor.
* Produces high seed yields but most seeds are hard and need scarification to ensure acceptable germination.

Use and Potential Problems
* Used for forage, green manure, and as cover crop especially on road banks.
* Is often a weed of pastures, hay fields, roadsides, and abandoned fields.
* Excellent erosion control plant for long-term stability and virtually no maintenance fertilizer.
* Provides some wildlife habitat.
* Quality and palatability of sericea lespedeza is limited by its high tannin content.
* Nonbloat causing legume primarily due to its tannin content.
* Will not die or significantly thin when growth is not removed annually.
* Considered as invasive.

Toxicity/Disorders
* None recorded.

Reference
* Sericea lespedeza [5,7,17,22,29−31,63,64]

Lespedeza, Shrub—*Lespedeza bicolor* Turcz. (Including *Lespedeza thunbergii* (DC.) Nakai)

Shrub lespedeza seeds

Lespedeza bicolor

Shrub lespedeza leaves (right, left, and center)

Shrub lespedeza plant

Leaf shape and arrangement—Leaves: Trifoliate, alternate, abundant dark green leaves.

Stolon/Rhizome/Roots: No stolons or rhizomes; fine stemmed.

Inflorescence: Rose-purple pea-like flowers found in both stem tips and in the upper leaf axils.

Shrub lespedeza leaves and flowers

Lespedeza, shrub—*Lespedeza bicolor* Turcz. (including *Lespedeza thunbergii* (DC.) Nakai)

Other common names: Shrub bush clover
Family: Fabaceae/Leguminosae
Life cycle: Warm season perennial
Native to: Asia

Distribution and Adaptation
- Found in the warm humid and transition zones.
- Requires well-drained soil and tolerates drought conditions.
- Grows on low fertility and acid sites.

Morphology/Growth pattern
- A loose, open shrub, or subshrub with a robust growth habit.
- The tallest of the lespedezas; commonly grows to 6.0 ft (1.8 m) but may reach 10.0 ft (3.0 m).

Use and Potential Problems
- Mostly used for wildlife and erosion control.
- Found in cultivated fields, waste areas, and roadsides
- Is considered as an invasive plant.

Toxicity/Disorders
- None recorded.

Reference
- Shrub lespedeza [5,7,17,22,29−31,63,64]

Medic, Black—*Medicago lupulina* L.

Black medic seeds with protruding hilum

Seed

- Size: 1.2—1.6 mm long, 0.9—1.2 mm wide
- Shape and texture: ovate, D- or kidney-shaped, compressed with a small notch on one long margin, the lower end of the notch protrudes, small round hilum in the notch; surface smooth, dull
- Color: yellow-brown with green and orange tinges; seedpod with swirled look
- Seeds per pound: 260,000

Black medic plant *Source: Lachlan Cranswick*

Black medic leaf with small projection *Source: Lachlan Cranswick*

Leaf shape and arrangement—Leaves: Leaflets wedge-shaped with a small spur or tooth at the tip; toothed leaf margins; the leaf's characteristics distinguishes black medics from other legumes especially hop clovers; **Petiole**: longer stalk (petiolule) of the central leaflet compared to lateral leaflets; **Stipules**: has spear-like stipules.

Stolon/Rhizome/Roots: No stolons or rhizomes; hairy stem; fibrous root systems.

Inflorescence: Flowers are small and yellow in globe-like clusters; clusters can be up to 13 mm across containing 10—50 flowers; flowers on long peduncle; as flower matures it forms a tightly coiled black seedpod;

Fruit: black, kidney-shaped, 1.5—3 mm long, contains a single seed.

Black medic seedhead/seed pod

Medic, black—*Medicago lupulina* L.

Other common names: Black clover, hop medic, yellow trefoil, hop medic, spotted burclover, blackseed
Family: Fabaceae/Leguminosae
Life cycle: Winter annual
Native to: Mediterranean region

Distribution and Adaptation
- Distributed over the warm humid and transition zones, and less prevalent in the cool humid region, especially in the northern most states.
- Grows naturally in pastures, meadows, roadsides, waste places, and turf.
- Adapted to a wide variety of soils with a good supply of lime in moist temperate climates.
- Not drought resistant.
- Remains green much later in spring than most other winter annuals.

Morphology/Growth Pattern
- Trailing stems arise from a tap root.
- Branched, slender, hairy, prostrate stems up to 32 in. (80 cm) long; shallow and thick, and are often difficult to pull from the ground.

Use and Potential Problems
- Low yielding.
- Well-known source of honey, the seeds have some value for birds and rodents. Also used for green fodder, green manure, and hay.
- Reseeds readily.
- The plant spreads easily and can form large colonies when left untouched.

Toxicity/Disorders
- Potential bloating agent.

Similar Species
- *Alfalfa* (*Medicago sativa* L.) a close relative of black medic is distinguished by its blue or purplish flowers and long black coiled seed pod. Black medic although has a coiled seed pod, it is easily distinguished from alfalfa by its leaf shape that are wedge-shaped with a small spur or tooth at the tip; the leaf's characteristics distinguishes black medics from other legumes especially *hop clovers* (*Trifolium campestre* Schreb.); is also often times confused with yellow woodsorrel (*Oxalis stricta* L.), however unlike black medic where the leaves are wedge-shaped and their central leaflets are on petiole, *yellow woodsorrel* leaves are heart-shaped and are unstalked.

Reference
- Black medic [5,7,12,17,22,29−31,44,47,63,64]

Trefoil, Birdsfoot—*Lotus corniculatus* L.

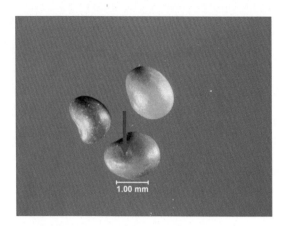

Birdsfoot trefoil seeds; seed with small dimple

Seed
- Size: 1.3–1.6 mm long, 1.0–1.3 mm wide
- Shape and texture: ovate with a small notch or dimple on one long side, compressed, hilum in notch is round; surface smooth, dull
- Color: brown or dark brown sometimes with small dark mottles
- Seed per pound: 375,000

Birdsfoot trefoil-5 leaflets, 3 at the end of petiole and 2 near the base

Birdsfoot trefoil plant *Source: Lachlan Cranswick*

Leaf shape and arrangement—Leaves: Five leaflets, 3 at the end of petiole and 2 near base like stipules; **Petiole**: short, diffused; **Stipules**: stipules reduced to a pair of darkish glandular spots.

Stolon/Rhizome/Roots: No stolons or rhizomes; short taproot.

Inflorescence: Yellow flowers 2.5–30 cm; umbellate heads; seed pods like a bird's foot; pods long, slender.

Birdsfoot trefoil flower *Source: Lachlan Cranswick*

Trefoil, birdsfoot—*Lotus corniculatus* L.

Other common names: Birdsfoot
Family: Fabaceae/Leguminosae
Life cycle: Cool-season perennial
Native to: Europe, North Africa, and part of Asia

Distribution and Adaptation

- Primarily grown in the cool humid region and the northern tier of the transition region and to a very limited extent at higher elevation of the transition zone.
- Grows on various types of soil, from clay to sandy loam.
- It will grow on infertile, or alkaline soils, on mine spoils and under saline and waterlogged conditions.
- It is more resistant than alfalfa to wet or waterlogged soils and tolerates low oxygen.
- Poor shade and drought tolerance.
- Most productive on fertile, moderately to well-drained soils having pH above 6.2.
- Long-day plant.

Morphology/Growth Pattern

- Its upright or spreading stem can reach 1.67—3.3 ft (51—102 cm) length.
- Taproot deep but not as deep as alfalfa; extensive branch roots in the upper soil layer.
- Many branched stems from a single crown.
- Growth cycles less distinct than alfalfa.
- Plants live for about 2—4 years; reseeding or reproduction of new shoots from existing plants necessary for persistence.
- Flowers later than most legumes because of long daylength requirement.
- Has less seedling vigor than alfalfa or red clover.

Use and Potential Problems

- Quality of forage excellent, equal to or better than alfalfa because of increased "bypass" protein and smaller stems.
- Does *not* cause bloat in cattle or sheep because of relatively high tannin content.
- Its attractive flowers and wide adaptation make it useful for highway slopes and medians for beautification, erosion control, and soil improvement.
- Regrowth is heavily dependent on leaf area following harvest/grazing thus, leave at least 3—4″ stubble.
- First cut should be at 25% bloom and subsequent cuts at 40-day intervals.
- Cultivars with prostrate type of growth will persist longer under close grazing than the upright types. Persistence and yield better if grazed rotationally instead of continuous.
- Trefoil is not as resistant as alfalfa to Fusarium-type diseases.

Toxicity/Disorders

- No potential bloating agent.

Reference

- Birdsfoot trefoil [5,7,17,18,22,29—31,57,63,64]

Vetch, Big Flower—*Vicia grandiflora* Scop.

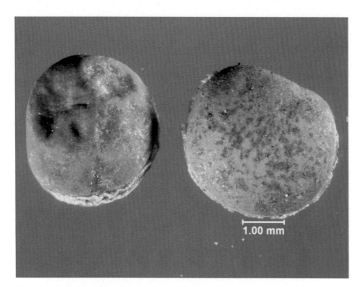

Big flower vetch seeds

Seed
- Size: 3.0–4.2 mm long, 3.0–3.2 mm wide
- Shape and texture: circular, compressed, rough surface, linear hilum
- Color: straw colored, may or may not be mottled with black

Big flower vetch plant

Big flower vetch stem and leaves

Leaf shape and arrangement—Leaves: with 3–7 pairs of thin, rather variable leaflets and a weak, mostly branched tendril; lower leaflets obovate (egg-shaped), upper leaflets oblong to linear.

Stolon/Rhizome/oots: No stolons or rhizomes; climbing stem, vine.

Inflorescence: Flowers usually in pairs in the axils of the upper leaves; white to pale yellow flowers. Similar to common vetch, except that big flower has two large yellow or white flowers in the axils of the leaves; seed pods slightly longer than hairy or common vetch.

Big flower vetch flowers

Vetch, big flower—*Vicia grandiflora* Scop.

Other common names: Large flowered vetch, white vetch
Family: Fabaceae/Leguminosae
Life cycle: Cool-season annual
Native to: Mediterranean of southern Europe

Distribution and Adaptation
- Found in all regions of the eastern United States.
- Grows well on sandy loams, clay-loams.
- Prefers pH of 5.5−7.5.
- Vetches in general are more tolerant to acid soil conditions than most legume crops.
- Have a relatively higher requirement for phosphorous than many other leguminous crops.

Morphology/Growth Pattern
- It has a climbing stem, viny.
- The stem bears leaves with pinnate leaflets and terminates in tendrils.
- The tendrils attach themselves to stems of associated field crops such as small grains

Use and Potential Problems
- Used for grazing and hay when mixed with winter annual grasses.
- Used as winter cover crop.

Toxicity/Disorders
- Cyanogenic glyosides.

Reference
- Big flower vetch [5,7,12,17,18,22,29−31,63,64,72,75]

Vetch, Common—*Vicia sativa* subsp. *sativa* L.

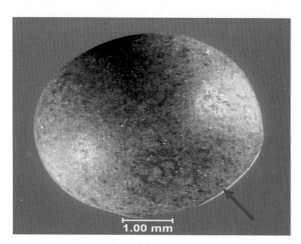

Common vetch seeds; distinct long hilum

Common vetch leaf with tendril *Source: NCSU*

Common vetch flower with leaf tip point in the middle

Seed

- Size: 4.5−6.0 mm long, 4.5−6.0 mm wide
- Shape and texture: irregularly round; smooth, dull, speckled, has distinct long hilum which looks like a fine white scratch on the surface
- Color: gray mottled with darker color
- Seeds per pound: 7016

Common vetch plant *Source: NCSU*

Leaf shape and arrangement—Leaves: Pinnately compound leaflets arranged opposite from each other; 4−8 pairs of oblong leaflets, with or without hairs; leaflets end in *terminal branching tendrils*; **Petiole/stipule**: Stipules normally occur on the bases of the petiole.

Stolon/Rhizome/Roots: With or without stolons, no rhizomes; with a simple or branched stem; fibrous root systems.

Inflorescence: Flowers mostly borne in pairs in the axils of the upper leaves, usually on short peduncle; petals violet-purple or rose, rarely white; the wing shorter and usually darker; fruits in flat pods.

Vetch, common—*Vicia sativa* subsp. *sativa* L.

Other common names: Winter tares, spring vetch
Family: Fabaceae /Leguminosae
Life cycle: Cool-season annual
Native to: Mediterranean region of southern Europe.

Distribution and Adaptation
- Found primarily in the warm humid and transition zones.
- Less winter hardy than hairy vetch but will tolerate temperatures of 10°F.
- Best adapted to well-drained loamy soil.

Morphology/Growth Pattern
- Stem can climb on upright associated crops.

Use and Potential Problems
- Can be used for grazing, hay production, or green manure.
- Is commonly used as cover crop.
- Also considered as a common weed of roadsides, landscapes, ornamentals, and row crops.

Toxicity/Disorders
- Cyanogenic glyosides.

Reference
- Common vetch [5,7,12,17,18,22,29−31,63,64,72,75]

Vetch, Crown—*Coronilla varia* L. = *Securigera varia* (L.) Lassen

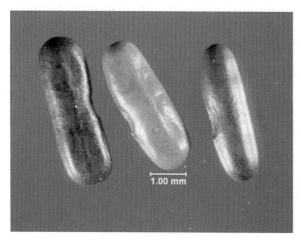

Crown vetch seed

Seed
- Size: 3.1—4.0 mm long, 1.0—1.3 mm wide
- Shape and texture: seed shaped like rod or cylindrical; surface smooth, glossy
- Color: seedpod yellow to mahogany, seed deep reddish red
- Seeds per pound: 110,000

Crown vetch plant

Crown vetch leaves

Leaf shape and arrangement—Leaves: The compound leaves consist of 15—25 pairs of oblong leaflets, leaf is "crowned" with a terminal leaflet, leaflets usually end in a small point, leaves are opposite, *no tendrils on leaf.*

Stolon/Rhizome/Roots: Crown vetch has very long rhizomes (up to 10 ft long), which allow the plant to spread rapidly; no stolon; fibrous root systems.

Inflorescence: Flower is arranged in a head-type, pink to blue; flower petals mostly pink or rose; flowers develop into narrow, long, and finger-like mature seed pods similar to birdsfoot trefoil.

Crown vetch flower *Source: Lachlan Cranswick*

Vetch, crown—*Coronilla varia* L. = *Securigera varia* (L.) Lassen

Other common names: Vetch
Family: Fabaceae/Leguminosae
Life cycle: Cool-season perennial
Native to: Central Europe

Distribution and Adaptation

- Found from the extreme northern part of the warm humid zone through the northern part of the cool humid region.
- It occurs along roadsides and other rights-of-way, in open fields and on gravel bars along streams.
- Established plants tolerant of moderately acid and infertile soils but best adapted to well-drained, fertile soils with a pH of 6.0 or above.
- Plant prefers open, sunny areas.

Morphology/Growth Pattern

- Perennial plant arising from thick branching rhizomes with trailing to weakly climbing stems.
- Can form large clumps from rhizomes.
- Spreads by seed and vegetatively by creeping rhizomes.

Use and Potential Problems

- Widely used for roadside beautification, erosion control, green manure, and mine spoil areas as well as other disturbed areas.
- May provide high-quality grazing for livestock, but not used in forage systems to any significant extent.
- Desirable characteristics include uniform and persistent ground cover
- Does not cause bloat in cattle.
- Poor seedling vigor and slow seedling growth rates are a limitation on getting good stands.
- It requires a very specific rhizobium for N fixation.

Toxicity/Disorders

- Cyanogenic glyosides.

Reference

- Crown vetch [7,12,29−31,63,64,75]

Vetch, Hairy—*Vicia villosa* Roth subsp. *villosa*

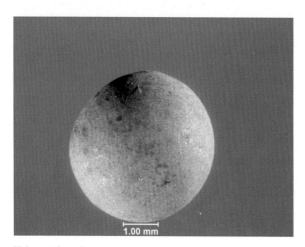

Hairy vetch seed

Pinnately compound leaf

Flowers on long racemes *Source: John Wright*

Seed
- Size: 2.7–4.9 × 2.3–4.9 × 2–3.8 mm (3–4.5 mm in diameter), usually smaller than common vetch
- Shape and texture: circular shape, indistinct hilum (if present short) surface smooth, dull, with a fine velvety sheen
- Color: dark reddish brown to dull greenish brown mottled with blackish brown
- Seeds per pound: 20,000

Hairy vetch plant

Leaf shape and arrangement—Leaves: Pinnately multifoliate terminating in branched tendrils, leaflets hairy (pubescent) (some cultivars nearly hairless), elliptical to lance shaped; the leaves and stem can be hairy or smooth; **Petiole**: short or absent; **Stipules**: pointed stipules.

Stolon/Rhizome/Roots: No stolon or rhizomes; taprooted.

Inflorescence: Petals bluish-violet and white, 20–30 flowers in long racemes; the densely clustered flowers are twisted on one-side, flowers borne on long stems or peduncles; seed pod elongated and compressed.

Vetch, hairy—*Vicia villosa* Roth subsp. villosa

Other common names: Sand vetch, winter vetch, woolypod, wooly vetch
Family: Fabaceae/Leguminosae
Life cycle: Cool-season annual
Native to: Europe and Asia

Distribution and Adaptation
- Found throughout the region but most often used in the warm humid and transition zones.
- Hairy vetch is the most winter hardy of the cultivated vetches.
- More tolerant of poorly drained soils than common vetch.
- Grows best on sandy or sandy loam soils.
- Tolerates soil pH ranging from 4.9 to 8.2

Morphology/Growth Pattern
- Climbing, prostrate, or trailing plants.
- Like common vetch and big flower vetch, the pinnate leaflets of hairy vetch terminate in much branched tendrils that attach themselves to stems of upright crops such as small grains or corn.
- Propagated by seed.

Use and Potential Problems
- Hairy vetch is used for grazing, hay, silage, but most often as a green manure cover crop. Often grown with wheat, oat, or rye. The grasses serve as companion crops and provide structural support to the legume.

Toxicity/Disorders
- Cyanogenic glyosides.

Reference
- Hairy vetch [5,7,12,17,22,29−31,46,47,63,64,72,74,75]

Hairy vetch	Common vetch	Crown vetch

Leaf arrangements   Leaf arrangements   Leaf arrangements

Flower   Flower   Flower

Seed   Seed   Seed

Grasses and Sedges

Common Grasses, Legumes and Forbs of the Eastern United States. DOI: https://doi.org/10.1016/B978-0-12-813951-6.00002-9

Bahiagrass—*Paspalum notatum* Flüggé

Bahiagrass seeds

Bahiagrass leaf blades *Source: NCSU*

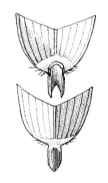

The collar region of bahiagrass

Bahiagrass seedhead *Source: NCSU*

Seed
- Size: 3.5 mm long, 2.5–2.75 mm wide
- Shape and texture: oval or ovate; glume coarsely wrinkled
- Color: light yellow
- Seeds per pound: for these plant species
- Bahiagrass: 273,000
- Bentgrass, Colonial, Bentgrass, Creeping: 6,000,000
- Bluegrass, Annual: 2,100,000
- Crabgrass Southern: 825,000
- Johnsongrass: 119,000

Rhizome and roots of bahiagrass

Leaf sheath and leaf blade—Vernation: Leaf rolled in the bud-shoot; **Leaf sheath**: flattened, sharply creased, rather glossy and usually not hairy; **Leaf blade**: leaf blade primarily basal, somewhat folded, usually sparsely hairy along edge toward base, tapering to a tip; **Collar**: broad; **Auricle**: Absent; **Ligule**: membranous, short, hairs on back.

Stolon/Rhizome/Roots: Has rhizomes; has stolons (very short and thick); plant with dense root system stolons; forms extensive root system.

Inflorescence: Seedhead with paired terminal branches, infrequently 1–3 additional branches below.

Bahiagrass—*Paspalum notatum* Flüggé

Other common name: None found
Family: Poaceae/Gramineae
Life cycle: Warm-season perennial
Native to: South America

Distribution and Adaptation
- Primarily in warm humid regions but in coastal plain and low elevations of the southern areas of the transition zone.
- Bahiagrass is the most utilized grass in Florida compared with any other single pasture species.
- Grows on a range of soil types from sandy to clayey, slightly acid, infertile soils.
- Produces moderate yield on low fertility soils and does not require high input of fertilizers.
- Not suitable for high pH soils because of pH-induced iron deficiency.
- It is drought tolerant.
- Does not tolerate shade.

Morphology/Growth Pattern
- Forms a dense sod/mat that can be very competitive toward associated species.
- Leaves are arranged close to the soil surface which makes it adaptable to close defoliation.
- The seedhead may reach 2−3 ft (60−90 cm) but leaves are basal and rarely reach 1.5 ft (45 cm).
- Bahiagrass has a deep root system extending 8−10 ft (2.4−3.0 m) deep in sandy textured soils.
- May be established from seed, sod, sprigs, or plugs.
- Seed germination is slow and seedlings are not vigorous nor competitive with weedy species.
- Reproduces by seed and rhizomes.

Use and Potential Problems
- Primarily used for pasture; when made into hay it is stemy.
- Used for sod, open turf, roadsides and other utility sites.
- It is difficult to mechanically harvest leaves of bahiagrass because they are so close to the soil surface.
- It is an invasive because of its seed production and its short, stout rhizomes.
- Repeated cultivation and herbicides are needed for its control.

Toxicity/Disorders
- None reported.

Reference
- Bahiagrass [2,7,13,29,32,49]

Barley—*Hordeum vulgare* L. subsp. *vulgare*

Barley seed

Seed
- Size: vary with variety
- Shape and texture: lemma and palea remain attached (unlike wheat); may have awn attached; lemma frequently wrinkled, six-rowed barley has a slightly twisted crease in seed and is longer and thinner than two-rowed barley
- Color: tan
- Seeds per pound: 13,000

Barley plant *Source: Carl Griffey and Wynse Brooks*

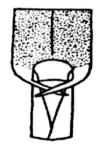

Collar region

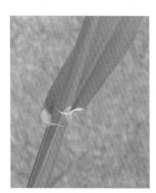

Barley collar region

Leaf sheath and leaf blade—Vernation: Leaf rolled in the bud, leaf blade flat, wide; **Leaf sheath**: split, overlapping, and glabrous; **Leaf blade**: 8−13 mm wide, rough above at least near tip, keeled and smooth below near base, margins smooth; **Collar**: smooth and narrow; **Auricle**: long, clasping (claw-like), glabrous; the barley auricles are larger than wheat or rye; **Ligule**: short or truncated.

Stolon/Rhizome/Roots: No stolons; no rhizomes; bunch growth habit; fibrous root system.

Inflorescence: Pike erect or slightly nodding.

Barley seedhead *Source: Wynse Brooks*

Barley—*Hordeum vulgare* L. subsp. *vulgare*

Other common name: None found
Family: Poaceae/Gramineae
Life cycle: Cool season, annual cereal grain
Native to: Southwest Asia

Distribution and Adaptation

- May be gown in the entire region with the exception of the very deep, warm humid zone.
- Prefers well drained soils; does not tolerate poor drainage.
- Sensitive to low pH, but tolerates salinity.
- It withstands drier, less fertile conditions than wheat.

Morphology/Growth Pattern

- Bunch type of growth habit; depending on variety can grow more than 2.5 ft (0.74 m).
- A single plant is composed of many tillers originating from the basal nodes. Earlier in the growing season, the tillers are mostly leaves, later they produce stems followed by seedheads.
- Depending on cultivar can grow up to 2−4 ft (0.6−1.2 m) tall.
- Two types of cultivated barley are differentiated by the development or lack of the lateral spikelets. In the two-rowed barley (seeds arranged in two rows), the lateral spikelets are sterile. In the six-rowed barley (six rows of seeds), however, the lateral as well as the central spikelets produce seeds.
- Where adapted, it is an early maturing, short season crop.
- Generally makes more winter growth than oat, but not as much as rye.

Use and Potential Problems

- Grazing, especially winter and mid spring grazing.
- Makes excellent silage harvested in boot to soft dough stage of growth.
- Used for winter cover crop.
- Reproduces by seed.

Toxicity/Disorders

- Not toxic, but mechanical injury from awns.

Reference

- Barley [7,8,25,26,32]

Barnyardgrass—*Echinochloa crus-galli* (L.) P. Beauv.

Barnyardgrass seeds

Seed

- Size: 3.0–3.3 mm long, 1.6–2.2 mm wide
- Shape and texture: strongly arched in side view, has extension on lemma with awns, if no awns, it comes to a sharp point; the dorsal side of the seed has light, longitudinal lines
- Color: tan
- Seed per pound:

Barnyardgrass plant (NCSU)

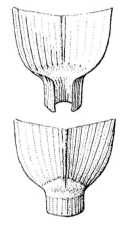

No auricle or ligule CIBA-GEIGY

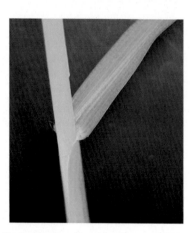

Barnyardgrass collar region *Source: NCSU*

Leaf sheath and leaf blade—Vernation: Leaf rolled in the bud; **Leaf sheath**: smooth, broad and long leaves; slightly flat with tuft of hair at base of plant; **Leaf blade**: smooth surfaced with rough margins; **Collar**: broad; **Auricle**: absent; **Ligule**: absent.

Stolon/Rhizome/Roots: No stolons; no rhizomes; stems thick and fleshy; fibrous root system.

Inflorescence: Branched, and covered with seeds that are tipped with short awns, panicle consists of nearly sessile one-sided racemes (in two to four rows), with long hairs at the nodes; spikelets are large, awned and covered with short stiff bristles.

Barnyardgrass seedhead (NCSU)

Barnyardgrass—*Echinochloa crus-galli* (L.) P. Beauv.

Other common names: Cockspur grass, barngrass, watergrass, duck millet, panic grass, summer grass, billion dollar-grass, cocksfoot panicum
Family: Poaceae/Gramineae
Life cycle: Warm-season annual
Native to: Europe

Distribution and Adaptation
- Grows throughout all zones.
- It appears late in the summer in unmanaged places, meadows, and roadsides.
- It grows on wide range of soils, but most found on sites with continuous moisture such as waterways, ditches, field edges, and low spots in pastures and cropland.

Morphology/Growth Pattern
- Grows erect and reaches a height of 2—4 ft (0.60—1.2 m).
- Prolific seed producer (as many as 40,000 seeds have been produced on a single plant).
- The tillers originate from the adventitious roots.
- Reproduces by seed.

Use and Potential Problems
- Seeds are important food for birds, especially waterfowl.
- Produces fair forage if grazed at the vegetative stage or cut for hay, but yields are usually low.
- In most situations it is considered an undesirable weedy grass in any landscape use.

Toxicity/Disorders
- Photosensitization agents.

Reference
- Barnyardgrass [12,15,16,30,32,59,64]

Bentgrass, Colonial—*Agrostis capillaris* L. (Formerly Known as *Agrostis tenuis* Sibth.)

Colonial bentgrass seeds

Seed

- Size: 1.5 mm long, 0.5 mm wide
- Shape and texture: lemmas lance-shaped; porous textured; basal hair lacking; palea relatively short, thin and wrinkled, apex notched, the points blunt
- Color: tan
- Seeds per pound:

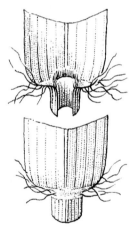

Collar region of Colonial bentgrass

Leaf sheath and leaf blade—Vernation: Leaf rolled in the bud; **Leaf sheath**: round, smooth, split with overlapping edges; **Leaf blade**: sharp pointed, upper surface distinctly ridged, slightly creased under surface, edge and upper surface rough, 1–3 mm wide; **Collar**: distinct, narrow, slanted with unequal sides; **Auricle**: absent; **Ligule**: membranous, entire or finely toothed, 0.4–1.2 mm long; blue-green in color.

Stolon/Rhizome/Roots: Short slender stolons and short rhizomes; fibrous root system.

Inflorescence: The flowers of creeping bentgrass are purplish, single flowered spikelets in a compact panicles.

Bentgrass, Creeping—*Agrostis stolonifera* L. var. *palustris* (Huds.) Farw. Previously Known as *Agrostis stolonifera* L.

Creeping bentgrass seeds

Seed
- Size: 1.5 mm long, 0.5 mm wide
- Shape and texture: lemmas ovate, the back humped up above the constricted base, callus thick and angled; paleas broad toward the top tapering abruptly to a minutely notched or rounded apex
- Color: straw to reddish in color
- Seeds per pound:

Collar region of Creeping bentgrass

Leaf sheath and leaf blade—Vernation: Leaf rolled in the bud; **Leaf sheath**: round, smooth, split; **Leaf blade**: flat distinctly ridged upper surface, slightly creased on lower surface and edges rough, 2−5 mm wide; **Collar**: distinct, narrow, slanted with unequal sides; **Auricle**: absent; **Ligule**: membranous, rounded or blunt, finely irregular-toothed or entire, slightly hairy on the back, 1−2 mm long; blue-green in color.

Stolon/Rhizome/Roots: Slender stolons and no rhizomes; fibrous root systems.

Inflorescence: Flowers of creeping bentgrass are purplish, single flowered spikelets in a compact panicles.

Collar region of creeping bentgrass *Source: NCSU*

Bentgrass, colonial—*Agrostis capillaris* L.

Bentgrass, creeping—*Agrostis stolonifera* L. var. *palustris* (Huds.) Farw. (previously known as *Agrostis palustris* Huds.)

Bentgrass, Colonial—*Agrostis tenuis* Sibth.

Other common names: None found
Family: Poaceae (Gramineae)
Life cycle: Cool-season, perennial
Native to: Europe

Distribution and Adaptation
- Grown throughout New England States and the Pacific Northwest, where bentgrass is well adapted, is used for lawn, golf courses, and athletic fields.
- Large genus with over 100 species, but only 4 are commonly used for turfgrass in the United States. Creeping bentgrass (bentgrasses) not well adapted to the southern climates, where used in the south, mostly used for putting greens where closely managed.
- Well adapted to the cool, humid environment (northeastern United States).

Morphology/Growth Pattern
- Creeping bentgrass spreads by stolons and forms a dense, fine textured sod thus forces out other weedy or turfgrass species.
- Grows up to 2.2–4.3 ft (0.66–1.3 m) tall.
- Reproduces by seed and stem pieces.

Use and Potential Problems
- Primarily known as a grass for golf course putting greens and fairways.
- The maintenance of bentgrass requires intensive management (specially in the south where not adapted well); the hot, humid summer can put bentgrass under stress thus specially during the summer months, watering, fertilization, and mowing should be closely managed.
- Due to poor stress tolerance and high maintenance requirement, not used for home lawns.

Toxicity/Disorders
- Ergot alkaloids.

Similar Species
- *Colonial bentgrass* (*A. tenuis* L.) differs from creeping bentgrass in that it has short stolons and short rhizomes while creeping bentgrass has only slender stolons.

Reference
- Colonial bentgrass [11,12,32,45,64,65]

Bermudagrass—*Cynodon dactylon* (L.) (Pers.) var. *dactylon*

Bermudagrass seeds; boat-shaped lemma

Seed
- Size: 2 mm long, 0.75−1 mm wide
- Shape and texture: triangular appearance, boat-shaped lemma
- Color: seed covering is light tan
- Seeds per pound: 1,800,000

Bermudagrass plant (NCSU)

Hairs at the base of the leaf blade

Bermudagrass seedhead

Leaf sheath and leaf blade—Vernation: Leaf folded in the bud-shoot; **Leaf sheath**: tuft of hair at base of leaf; is compressed, split with overlapping margins; **Leaf blade**: 2−5 mm wide, smooth to rough or short hairy above but mostly with a few hairs at the base; **Collar**: hairy on margins; **Auricle**: absent; **Ligule**: membranous; a fringe of hair.

Stolon/Rhizome/Roots: Has stolons; has rhizomes, fibrous root systems. Prostrate to semierect canopy.

Inflorescence: Three to six digitate spikes per head (*finger-like*) spikelets are one sided; look for triangular seed in seedhead.

Bermudagrass—*Cynodon dactylon* (L.) (Pers.) var. *dactylon*

Other common names: Wiregrass, couchgrass, devilgrasss, vinegrass
Family: Gramineae/Poaceae
Life cycle: Warm-season perennial
Native to: Southeast Africa

Distribution and Adaptation
- Found in the warm humid and transition zones, especially in the Coastal Plains and Piedmont regions.
- Adaptation is strongly related to soil drainage and winter survival.
- Tolerant of extended moisture stress and very short-term flooding, but does not persist on poorly drained soils.
- Grows well with adequate moisture and nutrients on moderately well-drained soil, whether acid or alkaline.
- Best adapted on sandy soils.

Morphology/Growth Pattern
- Morphology and growth patterns vary significantly depending on cultivar; can grow up to 0.33−1.1 ft (0.1−0.3 m).
- It tolerates close grazing or cutting because of leaf area near soil and multiple storage organs beneath grazing or cutting height.
- Grows best when mean temperatures are above 24°C (75.2°F), minimum growth when temperatures drop to 6−9°C (42.8−48.2°F) or below.
- Common bermudagrass is generally shorter than forage type hybrids and named seeded types.
- Common bermudagrass and the "seeded types" may be propagated by planting seed or vegetative sprigs, while hybrid cultivars must be established via planting rhizomes or stolons vegetatively.
- Generally, the hybrids have larger rhizomes, stems, and leaves than common and seeded types.
- Sod forming grass that spreads by rhizomes, stolons and by seed of the nonhybrids.

Use and Potential Problems
- Used for hay, grazing, turf, erosion control, and bioremediation.
- As a forage its nutritive value is only average, but its persistence and adaptability make it an excellent choice for livestock pastures.
- Is an excellent ground cover for erosion control and tolerates traffic and frequent defoliation.
- Is used as receiver crop of nutrients from confinement animal and municipal waste systems because on adapted soils it yields well.
- Common bermudagrass and cultivars of seed producing types produce enough hard seed to remain in the soil for many years, increasing the odds of becoming a serious weed.
- Hybrid bermudagrass is easier to control than seeded types because it spreads only via vegetative means.
- Reproduces by seed, stolons, and rhizomes.

Toxicity/Disorders
- Tryptophan.

Similar Species
- *Bermudagrass* is often confused with *large crabgrass* (*Digitaria sanguinalis* L.) mostly due to the morphological characteristics of their seedhead and creeping growth habit. Can be distinguished by noting *Bermudagrass* is a perennial plant with both stolons and rhizomes while *crabgrass* is annual with only stolons.

Reference
- Burmudagrass [1,4,7,9,12,16,23,32,45,64,65,72]

Bluegrass, Annual—*Poa annua* L.

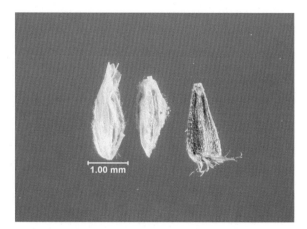

Annual bluegrass seeds

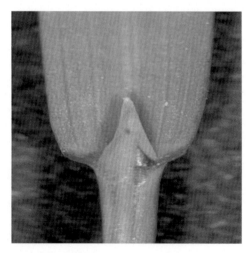

Annual bluegrass *Source: NCSU*

Annual bluegrass seedhead *Source: NCSU*

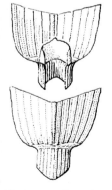

Ligule and collar region
of annual bluegrass

Seed (Annual bluegrass)
- Size: 2.5−3.0 mm long
- Shape and texture: has distinct fuzzy ridges giving a "hollow" appearance; plumper than Kentucky bluegrass, broader toward base
- Color: tan, caryopsis dark
- Seed per pound:

Annual bluegrass plant

Leaf sheath and leaf blade—Vernation: Leaf folded in the bud-shoot and is parallel sided; **Leaf sheath**: distinctly flattened; **Leaf blade**: with boat-shaped tip, slightly sharply creased, smooth; **Collar**: divided, distinct, smooth, and narrow to broad; **Auricle**: no auricle; **Ligule**: membranous, 1−3 mm long, rounded to acute.

Stolon/Rhizome/Roots: No stolons or rhizomes; the root system is fibrous and shallow.

Inflorescence: Open "pyramid-shaped" panicle; spikelets crowded, 3−6 flowered, about 4 mm long.

Bluegrass, annual—*Poa annua* L.

Other common names: Annual meadowgrass, wintergrass, annual spear-grass, dwarf spear-grass
Family: Poaceae (Gramineae)
Life cycle: Cool-season annual
Native to: Europe

Distribution and Adaptation
- Found throughout the world, mostly where excessively wet, compacted soils.
- Annual bluegrass is found in extremely overgrazed or high traffic areas where competition for other species is very low. Usually found on wide range of fine- to medium-textured soils.

Morphology/Growth Pattern
- Small, tufted to clumped, bright green; grows up to 11–12 in. (30 cm).
- Leaf blade smooth on both surfaces, with two distinct, clear lines located on one on each side of the mid-rib.
- Annual bluegrass has abundant basal leaves and is shorter than those of Kentucky bluegrass and exhibits wrinkles across the blade.
- Annual bluegrass may root at the stem node and forms a bunch type of growth.
- Annual bluegrass grows earlier in the season than Kentucky or roughstalk bluegrasses and produces seedheads starting in early spring.
- Reproduces seed under the closest grazing/cutting conditions and produces abundant amount of seeds.
- Reproduces by seed.

Use and Potential Problems
- Annual bluegrasses are considered weeds, especially in turf, vegetable and flower gardens, and landscapes.
- Annual bluegrass is too short (low yielding) and full of seedheads to be useful as a grazing plant; it is not a desired plant in forage systems.

Toxicity/Disorders
- None recorded.

Similar Species
- *Kentucky bluegrass* (*Poa pratensis* L.) is perennial and unlike *annual bluegrass Kentucky bluegrass* is rhizomatous, darker color, and produces fewer seedheads.

Reference
- Bluegrass [4,7,26,29,32,43,48,55,67,69]

Bluegrass, Kentucky—*Poa pratensis* L.

Kentucky bluegrass seeds

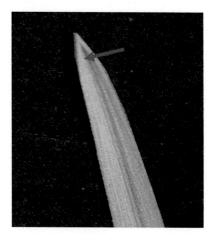

Boat-shaped tip need better shot *Source: NCSU*

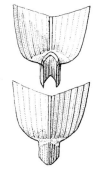

Ligule and collar region of Kentucky bluegrass

Seedhead "pyramid-shaped"

Seed (Kentucky bluegrass)
- Size: 3—3.8 mm in length, width of lateral plane 0.8 mm
- Shape and texture: elliptical; sharp, pointed lemma; numerous veins or nerves on lemma.
- Longer and darker than bermudagrass, broadest point of seed is in the center of caryopsis
- Color: tan
- Seed per pound: 2,200,000

Kentucky bluegrass plant *Source: NCSU*

Leaf sheath and leaf blade—Vernation: Leaf folded in the bud-shoot and is parallel sided; **Leaf sheath**: flattened, smooth, closed but split when mature; **Leaf blade**: usually v-shaped, sharply creased with boat-shaped tip and "clear track" as mid-vein; **Collar**: slightly divided by mid-rib, medium broad, may have fine hairs on edge; **Auricle**: no auricle; **Ligule**: membranous, very short, collar-like, 0.2—1 mm long, truncate.

Stolon/Rhizome/Roots: No stolons; rhizomatous, tufted and often spreading.

Inflorescence: Open panicle, "pyramid-shaped"; the lower most branches usually in a whorl of five panicle branches, normally one central long one, two shorter lateral ones and two short intermediate ones; spikelets crowded.

Bluegrass, Kentucky—*Poa pratensis* L.

Other common names: Smooth-stalked meadowgrass
Family: Poaceae (Gramineae)
Life cycle: Cool-season perennial
Native to: Europe

Distribution and Adaptation
- Commonly found in the cool humid and northern areas of the transition zone. 69 species of bluegrass are adapted and distributed in the United States.
- Kentucky bluegrass is a low-growing, persistent, and winter hardy species.
- Usually found on wide range of fine- to medium-textured soils. Kentucky and roughstalk are best adapted to moderately well-drained soils and grow best on soils of medium and high productivity.
- Temperature has the most influence on the distribution of the bluegrasses than any other climatic factors.

Morphology/Growth Pattern
- Kentucky bluegrass and roughstalk bluegrass are sod formers. Kentucky bluegrass can grow up to 1−2 ft (0.3−0.60 m).
- Leaves are arranged close to the soil surface which makes it adaptable to close defoliation.
- Annual bluegrass grows earlier in the season than Kentucky or roughstalk bluegrasses and produces seedheads starting in early spring.
- Seedhead formation is induced by cool, short days and regrowth following harvest or grazing of the perennial seed stalks is predominantly vegetative.
- All spread by seed but Kentucky bluegrass spreads by rhizomes and roughstalk spreads by stolons.

Use and Potential Problems
- The perennials are primarily used for grazing because leaves do not normally reach sufficient length for hay harvesting. Tolerates frequent and close grazing better than most cool-season grasses.
- Nutritive value and palatability are excellent for all of the bluegrasses.
- Kentucky bluegrass invades open spaces created when other species disappear due to overgrazing or low fertility.
- Reproduces by seed and rhizomes.

Toxicity/Disorders
- None recorded.

Reference
- Bluegrass [4,7,26,29,32,43,48,55,67,69]

Bluegrass, Roughstalk—*Poa trivialis* L.

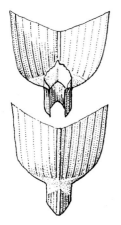

Ligule and collar region of rough-stalk bluegrass

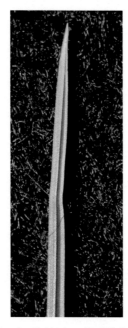

Roughstalk bluegrass leaf blade
Source: NCSU

Roughstalk bluegrass plant *Source: NCSU*

Leaf sheath and leaf blade—Vernation: Leaf folded in the bud-shoot and is parallel sided; **Leaf sheath**: flattened, sharply creased, usually rough, split part way only; **Leaf blade**: with boat-shaped tip, glossy, edges rough at least near tip, two distinct clear lines one on each side of the mid-rib; **Collar**: divided by mid-rib, broad, distinct; **Auricle**: no auricle; **Ligule**: membranous, 2−6 mm long, acute, sometimes ciliate.

Stolon/Rhizome/Roots: With stolons, no rhizomes; bunch/tufted rarely spreading.

Inflorescence: Open panicle, oblong, five panicle branches per node.

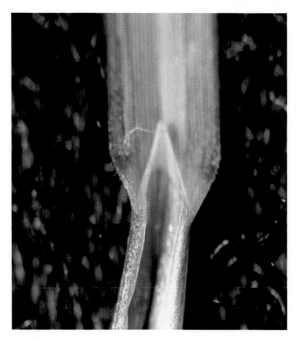

Roughstalk bluegrass collar region *Source: NCSU*

Bluegrass, roughstalk—*Poa trivialis* L.

Other common names: Rough-stalked meadowgrass, rough bluegrass
Life cycle: Cool-season perennial
Family: Poaceae (Gramineae)
Native to: Europe

Distribution and Adaptation
- Commonly found in the cool humid and northern areas of the transition zone.
- Roughstalk bluegrass is intolerant of full sun in midsummer unless irrigated; more shade tolerant than Kentucky bluegrass.
- Kentucky bluegrass and roughstalk are best adapted to moderately well-drained soils and grow best on soils of medium and high productivity.

Morphology/Growth Pattern
- Kentucky bluegrass and roughstalk bluegrass are sod formers; grow up to 3 ft (1 m).
- Roughstalk has more pale green, fine textured leaves than Kentucky bluegrass.
- Seed shatter much more than those of Kentucky bluegrass.
- All spread by seed but Kentucky bluegrass spreads by rhizomes and roughstalk spreads by stolons.

Use and Potential Problems
- Annual and roughstalk bluegrasses are considered weeds, especially in turf, vegetable and flower gardens, and landscapes.
- The perennials are primarily used for grazing because leaves do not normally reach sufficient length for hay harvesting.
- Nutritive value and palatability are excellent for all of the bluegrasses.

Toxicity/Disorders
- None recorded.

Similar Species
- All bluegrasses have boat-shaped leave tips and have no auricles. *Annual bluegrass* has no stolon or rhizomes, while Kentucky bluegrass and *roughstalk bluegrass* have rhizomes and stolons, respectively. The three bluegrasses can also be easily identified by the size of their ligules.

Reference
- Bluegrass [4,7,26,29,43,48,55,67,69]

Bluestem, Big—*Andropogon gerardii* (L.) Vitman

Big bluestem seeds

Big bluestem hairy stem and leaf sheath (Missouri)

Mature seedhead need more clear photo

Seed
- Size: glumes 7—9 mm in length, 1.5 mm in width
- Shape and texture: first glume has broad longitudinal groove on back
- Color: greenish brown in color
- Seeds per pound: 160,000

Big bluestem collar region C.E. Phillips, 1962

Leaf sheath and leaf blade—Vernation: Leaf rolled in the bud-shoot; young shoots somewhat flattened at the base; **Leaf sheath**: generally shorter than the internodes, purplish; base of lower sheath covered with silky hair; leaf sheath split; **Leaf blade**: hairy, flat, glabrous on underside, rough margins and long; **Collar**: narrow to broad and sometimes divided; **Auricle**: absent; **Ligule**: membranous, sometimes fringed.

Stolon/Rhizome/Roots: No stolons, may or may not have rhizomes; fibrous roots; bunch-type habit; stem solid.

Inflorescence: Terminal panicle that is usually composed of two or three spike-like racemes; commonly referred to as "turkey foot" shaped; usually purplish, but can be yellowish; rachis straight, the joints and pedicels hairy on the sides and sessile spikelets at the base; the lemma of the spikelet has an awn that is bent and twisted basally.

Bluestem, big—*Andropogon gerardii* (L.) Vitman

Other common names: Turkey-foot, beard grass
Family: Poaceae (Gramineae)
Life cycle: Warm-season perennial
Native to: North America

Distribution and Adaptation

- Found primarily in the western areas of the cool humid and transition zones, however may be found anywhere in the eastern United States.
- Best adapted to deep, fertile, and moist soils but also grows well on shallow, gravelly ridges of pastures, roadsides, field borders, and buffer areas.
- Can produce relatively well on low fertility and moisture stressed conditions as compared to cool-season grasses.

Morphology/Growth Pattern

- It is a tall upright plant reaching up to 6.0 ft (1.8 m) at maturity.
- May form large bunches with extensive root system reaching 5.0 ft (1.5 m) on some soils.
- Leaves are green most of the summer and in the fall the stem/leaf sheath bluish-purple color.
- Seedlings are shade tolerant compared to most grasses, but often cannot compete with vigorous growth of summer annual grass weeds.
- Matures about 2–3 weeks later in the growing season than switchgrass.
- Reproduces by seeds and short rhizomes.

Use and Potential Problems

- Used primarily for grazing but also makes good hay.
- Has high palatability and nutritive value for most types of grazing animals.
- Provides good habitat for small game and birds because of its somewhat clumpy cover and seed production.
- Biggest limitation on widespread use on farms is related to establishment challenges. The chaffy (long awns) and impure nature of the seeds make it difficult to distribute using conventional planting equipment.

Toxicity/Disorders

- Alkaloids produced by fungal endophytes.

Reference

- Big bluestem [7,12,28,32,34,39,54,55]

Bluestem, Caucasian—*Bothriochloa bladhii* (Retz) S.T. Blake

Caucasian bluestem seeds

Caucasian bluestem hairy collar region

Caucasian bluestem seedhead

Seed
- Size: 2 mm
- Shape and texture: long and hairy
- Color: reddish brown
- Seeds per pound: 900,000

Caucasian bluestem plant

Leaf sheath and leaf blade—Vernation: Leaf rolled in bud-shoot; **Leaf sheath**: split, hairy or glabrous; **Leaf blade**: flat and narrow, 5−38 cm long, glabrous, prominent midrib, tapers to a fine point; **Collar**: medium broad; **Auricle**: absent; **Ligule**: hairy.

Stolon/rhizome/root system: No stolons; short rhizomes; fibrous, relatively shallow root system.

Inflorescence: Elongated panicle, 6−15 cm long, pale green to purplish, with the axis longer than the branches; the branches are slender, and whorled.

Bluestem, Caucasian—*Bothriochloa bladhii* (Retz) S.T. Blake

Other common names: Australian bluestem, Australian beardgrass
Family: Poaceae (Gramineae)
Life cycle: Warm-season perennial
Native to: Soviet Union, Australia, and South Asia

Distribution and Adaptation
- Adapted to the southern cool humid region and the transition zone with limited use in the southern parts of the warm humid region.
- Part of the group known as Old World bluestems (*Bothriochloa*, *Capillipedium*, and *Dichanthium*). It is not related to native bluestems.
- Tolerates moisture stress but will respond to moisture.
- Grows on wide range of soils, especially medium- to fine-textured soils and produces relatively well on shallow, rocky soils.
- Not well adapted to extremely sandy soils, wetlands and saline soils.
- Grows on low pH soils (as low as 5.2).

Morphology/Growth Pattern
- Is a bunch grass; can grow to height of 2−4 ft (0.6−1.2 m).
- Spring growth begins several weeks after switchgrass or big bluestem.
- Regrowth of Caucasian bluestem will develop seedheads after each harvest or grazing
- Spreads by short rhizomes and seeds.

Use and Potential Problems
- Used primarily for pasture and too much lesser extent for hay.
- Palatability and nutritive value are only average and significantly less than switchgrass or Indiangrass.
- Chaffy and fluffy seeds are difficult to plant in conventional equipment.
- Seedling vigor is only fair.
- Seeds are easily spread by wind and animals and may contaminate other areas.
- Provides poor habitat for wildlife.
- Persistence is highly related to providing sufficient leaf area on the canopy during the autumn period so that plants can accumulate reserve energy and root development for winter survival.

Toxicity/Disorders
- Alkaloids produced by fungal endophytes.

Reference
- Caucasian bluestem [4,7,12,32,74]

Bluestem, Little—*Schizachyrium scoparium* (Nutt.) Engelm.

Little bluestem seed

Seed
- Size: spikelet 6–8 mm long, 1 mm wide
- Shape and texture: pedicel and rachis segment stout, shorter than the glumes, flattened below and expanded toward the top
- Color: glumes yellowish or brownish
- Seed per pound: 240,000

Leaf sheath and leaf blade—Vernation: Leaf folded in the bud; **Leaf sheath**: split, keeled; **Leaf blade**: flat, blade tapering to tip, bluish basal shoots; **Collar**: divided; **Auricle**: large, horn-like; **Ligule**: hairy, membranous.

Stolon/rhizome/root system: No stolons, but has short rhizomes; fibrous extensive root system.

Inflorescence: Curved, slender racemes, one raceme per peduncle, hairs protruding from the spikelets; awned, sheath sometimes enclosed peduncle.

Zigzag rachis; fuzzy hair protruding from the spikelets http://www.lib.ksu.edu/wildflower/grass/littleblue10.jpg

Bluestem, little—*Schizachyrium scoparium* (Nutt.) Engelm.

Other common names: Prairie beardgrass, broom beardgrass, old man's beard
Family: Poaceae (Gramineae)
Life cycle: Warm-season, perennial
Native to: North America

Distribution and Adaptation
- Distributed mainly in cool humid, transition, and northern warm humid zones.
- Grows on a wide range of soil conditions but grows best on calcareous soils derived from limestone.
- Typically found on dry, upland sites and along hillsides and ridges, along roadsides, and utility rights-of-way.
- It is found on unmanaged and poorly performing sites where it provides marginally good soil cover.
- Is drought resistant due to its extensive, dense root system.
- Is not shade tolerant.

Morphology/Growth Pattern
- It has bunch growth habit with seedheads reaching 2–4 ft tall (0.61–1.2 m).
- Spreads by seed, tillers, and short rhizomes.

Use and Potential Problems
- As a forage it is primarily grazed, but may be used of hay to much less extent. However, some consider it undesirable for grazing due to its clumpy growth habit.

Toxicity/Disorders
- Alkaloids produced by fungal endophytes.

Reference
- Little bluestem [12,28,32,39,48]

Bromegrass, Downy—*Bromus tectorum* L.

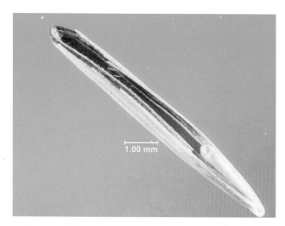

Downy brome seed

Seed

- Size: 8.0−12.0 mm long, 0.9−1.1 mm wide
- Shape and texture: lanceolate, tapers to the ends; long and slender with distinct long curved awn, sway-backed appearance, hairy
- Color: grayish-yellow to purplish
- Seeds per pound:

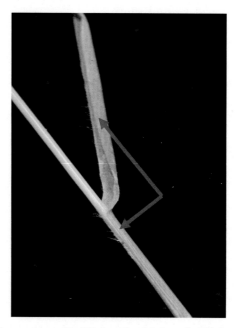

Downy brome stem and leaf blade with soft hairs on leaf blade and margins

Downy brome seedhead

Leaf blade and leaf sheath—Vernation: Leaf rolled in the bud; **Leaf sheath**: round, keeled, softly hairy, with prominent pinkish veins, split part way only; **Leaf blade**: 5.1−12.7 cm long, 0.42−0.84 cm. wide, flat, soft hairs on both sides and on margin, young leaves usually twisted; **Collar**: hairy, narrow, divided; **Auricle**: absent; **Ligule**: membranous, jagged 1 mm long.

Stolon/rhizome/root system: No stolons; short rhizomes; fibrous roots.

Inflorescence: Open panicle, soft and drooping spikelets on slender stalks; at maturity spikelets reddish-purple.

Bromegrass, downy—*Bromus tectorum* L.

Other common names: Downy chess, slender chess, wild oat, military grass, cheatgrass, cheatgrass brome, downy bromegrass, early grass, wild brome, drooping brome, wall brome, hobograss
Family: Poaceae (Gramineae)
Native to: Europe
Life cycle: Winter annual

Distribution and Adaptation
- Found mostly in the cool and warm humid regions.
- Found on dry, sandy, or gravelly soils in pastures, roadsides, turf, winter grains, and other row crops.

Morphology/Growth Pattern
- Plant grows up to 2.0 ft (0.61 m) in height depending on habitat.
- Flowers in late spring and dies.

Use and Potential Problems
- Usually considered an undesirable plant because it is competitive with other species and is of relatively low nutritive value, especially after stem elongation.
- Provides food for birds, turkeys, deer, and prairie dogs.
- The seed is frequently infected with smut.
- Frequent and close defoliation in early spring can provide some control.

Toxicity/Disorders
- Not toxin but mechanical injury from awns.

Similar Species
- *Downy brome* is easily confused with *cheat* (*Bromus secalinus* L.); however, cheat is smooth with occasional hair on the blade and lower sheath and the awns of cheat are shorter than those of downy brome.

Reference
- Downy bromegrass [23,28,32,39,48]

Bromegrass, Smooth—*Bromus inermis* Leyss. subsp. *inermis*

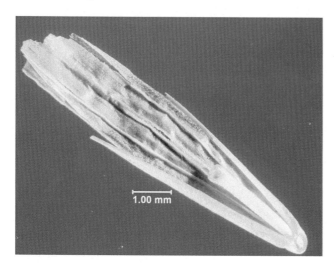

Smooth bromegrass seed

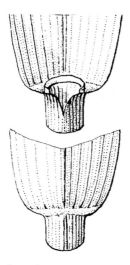

Smooth bromegrass collar region

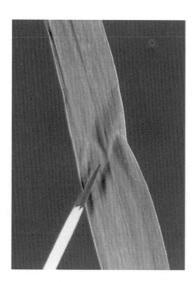

"M" or "W"

Seed

- Size: lemma 9—12 mm long, 2 mm wide
- Shape and structure: "snowshoe"; flat with distinct veins showing on lemma; lemma and palea appear smooth and membranous
- Color: brown to straw-colored
- Seeds per pound: 137,000

Smooth bromegrass plant

Leaf sheath and leaf blade—Vernation: Leaf rolled in the bud-shoot; **Leaf sheath**: round, smooth, frequently short hair on lower sheath, closed near top; **Leaf blade**: often constricted to form an "M" or "W" (not always present); **Collar**: medium broad to broad, divided; **Auricle**: absent; **Ligule**: round-edged, membranous, mostly less than 1 mm long, truncated to rounded.

Stolon/rhizome/root system: No stolons; extensive rhizomes; extensive root system.

Inflorescence: Erect, dense, open panicle from 6.1 to 17 cm long bearing 6- to 11-flowered spikelets.

Bromegrass, smooth—*Bromus inermis* Leyss. subsp. *inermis*

Other common names: Skinner's Gold, brome, Austrian brome, Hungarian brome, Russian brome bromegrass
Family: Poaceae (Gramineae)
Life cycle: Cool-season perennial
Native to: Central and Northern Europe and temperate Asia, extending to China

Distribution and Adaptation
- Primarily found in the western areas of the cool humid zone and too much lesser extent in the NE and northern area of the transition zone.
- Suited to silt or clay soils, deep loams, but also does well on light sandy soils, and well-drained soils.
- Survives long-term moisture stress and extremes in temperature.
- Not adapted to saline or alkali soils.

Morphology/Growth Pattern
- Stem/culms are erect and flowering plants may reach 4 ft tall (1.2 m).
- Deep rooted, but has a mass of roots in the surface 12 in. (30.1 cm) of the soil.
- Like most cool season grasses, it requires a period of growth under cool short days before exposed to warm and long days for the first seedhead production. However, unlike many other cool-season grasses, smooth bromegrass produces seedheads after subsequent cuttings.
- Sod former, spreads vegetatively by rhizomes and to lesser extent by seeds.

Use and Potential Problems
- Used for hay and pasture.
- Relatively high nutritive value and palatable to animals.
- Is used for soil stabilization/ erosion control.
- Intolerant of close and frequent defoliation or heavy traffic because its growing point elevates following each harvest or grazing.
- Disadvantage is its slow recovery after harvest or grazing.

Toxicity/Disorders
- Ergot alkaloids.

Similar Species
- *Nodding brome* (*Bromus anomalus* Rupr.) resembles smooth bromegrass. However, nodding brome can be easily distinguished from smooth bromegrass by its drooping panicle, and fibrous roots and lack of rhizomes.

Reference
- Smooth bromegrass [7,12,16,28,32,54,56]

Broomsedge—*Andropogon vinginicus* L. Leyss.

Bromesedge seed

Bromesedge leaf blades and collar regions

Bromesedge collar region

Bromesedge seedhead

Seed
- Size: 3–3.5 mm long and 0.5 mm wide
- Shape and structure: hairy, with bends
- Color: brown to straw-colored
- Seed per pound: 800,000

Bromesedge plant

Leaf sheath and leaf blade—Vernation: Leaf rolled in the bud-shoot; **Leaf sheath**: split, very flattened, sharply creased, long hairs on edges; **Leaf blade**: flat to v-shaped, smooth; **Collar**: divided by mid-vein, narrow, hairy at the edges; **Auricle**: absent; **Ligule**: short to medium, yellow-brown, membranous, truncate, white-fringed at edge.

Stolon/Rhizome/Roots: No stolons, very short, *almost unnoticeable rhizomes; fibrous root system.*

Inflorescence: Inflorescence in raceme, slender and narrow with two to four seeding stalks.

Broomsedge—*Andropogon vinginicus* L. Leyss.

Other common names: Yellow bluestem, whisky grass, beard-grass, broomstraw, sagegrass
Family: Poaceae (Gramineae)
Life cycle: Warm-season perennial
Native to: North America, Central America, and West Indies

Distribution and Adaptation
- Found in the entire region.
- Mostly found in open, sunny locations on low-fertility and droughty soils.
- Common invader of pastures, hay fields, abandoned fields, coal strip mines, and quarries.
- Often is unnoticed until it matures into a reddish-brown clump of high fiber stems that remain throughout the winter.

Morphology/Growth Pattern
- A bunch grass that grows up to 3.0 ft (0.91 m) height.
- Dense fibrous root system from short rhizomes.
- Seeds are dispersed by wind and clumps may expand by the short rhizomes.

Use and Potential Problems
- It has relatively little forage value beyond the early vegetative stage because of low animal consumption and high fiber composition.
- Considered a weed but is sometimes utilized when in very young stages of growth.
- Its main benefit to wildlife is for cover during winter months because it remains upright through the winter and subsequent spring.
- Is a common plant species associated with quail nests.
- Small birds remove and eat seeds from the flowering stalks in the winter when the seeds of other plants are unavailable.
- To assist in control broomsedge bluestem adjust soil pH to >6.0 and maintain high soil phosphorus.

Toxicity/Disorders
- Alkaloids produced by fungal endophytes.

Reference
- Broomesedge [12,23,32,39,43,45,54,69]

Buffalograss—*Buchloe dactyloides* (Nutt.) Engelm.

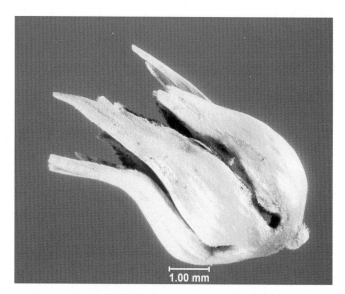

Buffalograss seed

Seed (bur)
- Size:
- Shape and texture: 3−5 parted lump with holes each segment
- Color: tan
- Seeds per pound: 40,000−55,000

Male plant

Leaf sheath and leaf blade—Vernation: Leaf rolled in bud-shoot; **Leaf sheath**: rounded, smooth, shorter than internodes; **Leaf blade**: 1.27 cm wide and 7.62−15.24 cm long sometime the leave blade curls; **Collar**: broad; **Auricle**: none; **Ligule**: row of shorter hairs.

Stolon/Rhizome/Roots: Has stolons; no rhizomes; fibrous roots.

Inflorescence: Spike type of seedhead, female flowers are in sessile burs, partly hidden between leaves; each bur contains one to four seeds; male flowers are in two or three short spikes on slender, erect stems.

Female plant with bur

Buffalograss—*Buchloe dactyloides* (Nutt.) Engelm.

Other common names: None known
Family: Poaceae (Gramineae)
Life cycle: Warm-season, perennial
Native to: United States

Distribution and Adaptation
- Distributed throughout short-grass prairie region of Great Plains.
- Adapted best to loamy clay soils; dominant on sites that are intermittently wet and dry, withstands prolonged summer droughts.

Morphology/Growth Pattern
- Grows up to 4−6 in. (10.1−15.2 cm) tall.
- Male and female plants grow separately (in separate colonies). The female plants produce seed in small hard burs that are usually nestled close to the ground among the leaves. The male plants produce the pollen and have two or three comb-like seedheads.
- Reproduces by seed and stolons.

Use and Potential Problems
- Used for turf and erosion control in the humid and arid regions and for grazing in arid regions.
- Buffalograss withstands prolonged close defoliation better than most native grasses; it has many leaves near soil surface.

Toxicity/Disorders
- None recorded.

Reference
- Buffalograss [29,30,31,32,57,58,71]

Canarygrass, Reed—*Phalaris arundinacea* L.

Reed canarygrass seeds

Reed canarygrass collar region

Reed canarygrass seedhead

Seed
- Size: lemmas 3—3.5 mm long, 1 mm wide
- Shape and texture: very sharply pointed; two hairy bristles at base of seed
- Color: grayish brown, very glossy
- Seeds per pound: 533,000 seeds/lb

Reed canarygrass plant

Leaf sheath and leaf blade—Vernation: Leaf rolled in the bud-shoot; **Leaf sheath**: round, smooth, split with overlapping hyaline margins; leaf sheaths transparent along the edges; **Leaf blade**: 10—20 mm wide; smooth to slightly rough and sometimes hairy near base above; margins usually rough and rarely short hairs near base; **Collar**: narrow to broad; may be divided; **Auricle**: absent; **Ligule**: membranous, 3—7 mm long; rounded to acute, occasionally toothed, may be hairy on back.

Stolon/Rhizome/Roots: No stolons, with rhizomes, deep rooted.

Inflorescence: Compressed, spikelets long, pointed; one floret per spikelet; spikelets on all sides of central axis; short first panicle branch compared with orchardgrass which is much longer.

Canarygrass, reed—*Phalaris arundinacea* L.

Other common names: Reed, canarygrass
Family: Poaceae (Gramineae)
Life cycle: Cool-season perennial
Native to: Europe

Distribution and Adaptation
- Best adapted to and found in the cool humid region. Limited use in the transition zone and rarely found in the warm humid zone, with the exception of higher elevations.
- It tolerates wet and dry extremes; persist on soils too poorly drained for most other forages.
- Moderately adapted to saline soils but tolerates a pH range of 4.9−8.2.
- It is intolerant of shade.

Morphology/Growth Pattern
- The tall, course, often hollow stems with some reddish coloration near the top of the stem; may reach 3−6 ft (0.91−1.83 m) in height.
- Will also produce roots and shoots from freshly cut, well-jointed reproductive stems.
- All new growth produces stems (culms) with elevated growing point (terminal meristem) therefore, frequent and close cutting/grazing may kill the tillers before new shoots and reserve energy is sufficient to support regrowth. This characteristic is similar to smooth bromegrass.
- Poor establishment rates sometimes due to low seed vigor and germination.
- Reproduces by seed and rhizomes and is a strong sod former.

Use and Potential Problems
- Used for pasture, hay, and silage.
- The major disadvantage of this grass for livestock feed is attributed to low palatability related to high alkaloid content.
- Where well adapted it has been used as a receiver crop for effluent from municipalities and confinement animal operations.
- Used in conservation programs for the protection of waterways riparian areas, and stream channel banks. May become invasive and threatening in some wetland ecosystems because of its competitive attributes.
- Sheep tend to be more adversely affected by the alkaloid composition than other domestic animals.

Toxicity/Disorders
- Alkaloids.

Reference
- Reed canarygrass [4,7,12,29−32,54,55,63]

Centipedegrass—*Eremochloa ophiuroides* (Munro) Hack

Centipedegrass seeds

Collar region of Centipedegrass

Centipedegrass from (NCSU)

Seed
- Size: 2 mm long, 1–1.5 mm wide
- Shape and texture: caryopsis oval, flattened dorsoventrally
- Color: hulled seed reddish brown
- Seeds per pound:

Centipedegrass plant (NCSU)

Leaf sheath and leaf blade—Vernation: Leaf folded in the bud; **Leaf sheath**: strongly compressed/flattened, grayish tufts of hairs at throat, edges overlapping; **Leaf blade**: flattened, short, sharply creased, hairs along ages at the base, 3–5 mm wide; **Collar**: continuous, broad constricted by fused crease, hairs tufted at the lower edge; **Auricle**: absent; **Ligule**: short, membranous with fine hairs.

Stolon/Rhizome/Roots: With slender, branching stolons that root at the node; no rhizomes.

Inflorescence: Inflorescence is a single spike-like raceme that range from 7.6 to 17.8 cm in length; purplish racemes somewhat flattened; spikelets arranged in two rows.

Centipedegrass—*Eremochloa ophiuroides* (Munro) Hack

Other common names: Lazy man's grass, poor man's grass
Family: Poaceae (Gramineae)
Life cycle: Warm-season, perennial
Native to: Asia and Europe

Distribution and Adaptation
- Native to China and southeast Asia.
- In the United States, widely grown in the southeast from North Carolina to Florida and westward along the Gulf Coast states to Texas.
- The grass prefers full sun but will tolerate some shade and is heat tolerant.
- Adapted to sandy, acidic soils, tolerates low fertility and requires minimum amount of maintenance.
- Centipedegrass is somewhat more cold tolerant than St. Augustine grass.

Morphology/Growth Pattern
- Centipedegrass has leaf blades 15–30 mm long, 2–4 mm wide, flat, rounded at the base, leaves are petioled, sparsely haired more so along the margins and at the bases of the stem; leaf sheath is overlapping, compressed and is pubescent at the base of the leaf sheath; inflorescence a spike-like raceme, purplish in color.
- Centipedegrass is a coarse-textured grass that spreads by stolons.
- Due to the stolons has a creeping growth habit with short upright stem.
- Forms a dense turf growth at a relatively low rate.
- Dormant during frost period.
- Propagated by seed, plugs, sprigs, or stolons.

Use and Potential Problems
- Used primarily for golf courses, lawns, parks, roughs and utility turf.
- Similar to St. Augustine grass, centipedegrass does not tolerate heavy traffic and not used for athletic fields.
- The stolons have a high lignin content and therefore the rate of decomposition is slow and thus develops a thatch layer.
- Centipedegrass requires less mowing than bermudagrass or St. Augustine grasses; due to its low maintenance requirement has been described as "lazy man's grass."
- In mild climate, remains green throughout the winter season; in hard freeze however, leaves and young stolons will die.
- Due to its low nutritive value, has little use for livestock feed; but useful for erosion control.
- Highly susceptible to damage by nematodes. Susceptibility to nematodes makes centipedegrass less suitable for use on some deep sandy soils.

Toxicity/Disorders
- None recorded.

Reference
- Centipedegrass [11,32,45,64,65]

Cheat—*Bromus secalinus* L.

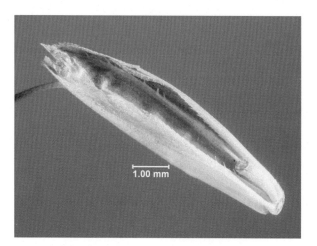

Cheat seed

Seed
- Size: 6.5−8.0 mm long, 1.5−2 mm wide
- Shape and texture: elongated, narrowly elliptical, horseshoe-shaped in cross section, awnless or with short awn curved deep, course teeth on edge of palea overlapping lemma; a fringe of stiff hairs on the margins of the inner bract (palea)
- Color: yellow to gray
- Seeds per pound:

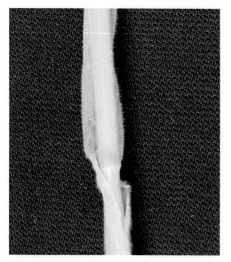

Hairy leaf sheath and stem

Leaf sheath and leaf blade—Vernation: Leaf rolled in the bud; **Leaf sheath**: round, closed, smooth; **Leaf blade**: smooth 7.6−25.4 cm long, 0.25−0.64 wide, sharp-pointed, long white hairs above, short white hairs below; **Collar**: broad, divided; **Auricle**: absent; **Ligule**: membranous, 0.31 cm long, toothed.

Stolon/Rhizome/Roots: No stolons; no rhizomes; fibrous.

Inflorescence: Much branched open, dropping panicle; has numerous spikelets, each having 6−11 florets; spikelets suspended on long peduncles.

Long peduncle

Cheat—*Bromus secalinus* L.

Other common names: Chess, sofeet chess, false wheat, smooth rye brome, wheat-thief, Willard's bromegrass, cockgrass
Family: Poaceae (Gramineae)
Life cycle: Winter annual
Native to: Asia and Europe

Distribution and Adaptation
- Found throughout the United States, but mostly in the cool humid regions.

Morphology/Growth Pattern
- The plant grows 2−4 ft (0.61−1.2 m) high.
- Stem smooth, unbranched stems, leaves, alternate, linear, and smooth.
- Reproduces by seed.

Use and Potential Problems
- Considered a noxious weed in many states
- A common found in grain crops, forage grasses, and roadsides.

Toxicity/Disorders
- Not toxin but mechanical injury from awns.

Reference
- Cheat [16,23,32,36,44,47,72,76]

Crabgrass, Large—*Digitaria sanguinalis* (L.) Scop.; Crabgrass, Smooth—*Digitaria ischaemum* (Schreb.) Muhl.; Crabgrass, Southern—*Digitaria ciliaris* (Retz.) Koeler

Large crabgrass seeds

Smooth crabgrass seeds

Large crabgrass seedhead *Source: NCSU*

Seed
- Size: 2.5–3.5 mm long, 0.9–1.1 mm wide
- Shape and texture: flattened on one side, round to oval on the other side; striped and hairy, three veins on flat side of seed
- Color: tan colored
- Seed per pound:

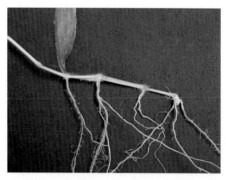

Large crabgrass, stolons, and root at the node *Source: NCSU*

Leaf sheath and leaf blade—Vernation: Leaf rolled in the bud-shoot (for both smooth and large crabgrass); **Leaf sheath**: distinctly flattened, very rough, split overlapping margins (*large crabgrass*), smooth or occasionally sparsely hairy (*smooth crabgrass*); **Leaf blade**: *large crabgrass* is flat, covered with short silky hairs on both surfaces of the leaf blade; *smooth crabgrass* similar to large crabgrass but has few hairs on the blade and sheath and is smaller in size; **Collar**: large crabgrass has medium broad to broad, mostly divided, hairy at least on margins; *smooth crabgrass* narrow to broad, often divided, may be hairy on margins; **Auricle**: absent; **Ligule**: membranous, toothed, acute, stiff, and papery; white hairs on the edges of the leaf but there are none on the ligule.

Stolon/Rhizome/Roots: Has stolons, no rhizomes; roots are fibrous with adventitious roots from the node of elongated tiller; large crabgrass roots at the stem nodes while smooth crabgrass does not.

Inflorescence: Flowers are borne in a raceme with 3–13 (fewer for smooth crabgrass) purplish finger-like spikes up to 15.2 cm long; they occur at end of stout stalks.

Large Crabgrass and Smooth Crabgrass Characteristics

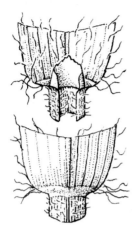

Large crabgrass

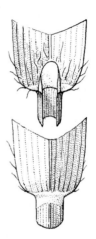

Smooth crabgrass

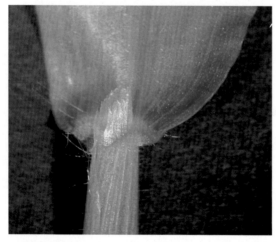

Large crabgrass ligule

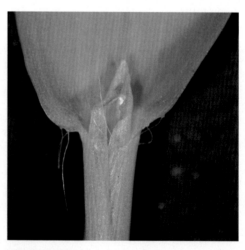

Smooth crabgrass ligule

Large crabgrass, hairy leaves and stem *Source: NCSU*

Smooth crabgrass, smooth leaves and stem *Source: NCSU*

Crabgrass, large—*Digitaria sanguinalis* (L.) Scop.

Crabgrass, smooth—*Digitaria ischaemum* (Schreb.) Muhl.

Crabgrass, southern—*Digitaria ciliaris* (Retz.) Koeler

Other common names: For large crabgrass: Finger grass, hairy crabgrass
Family: Poaceae (Gramineae)
Native to: Southern Africa
Life cycle: Summer annual

Distribution and Adaptation
- Distributed throughout the United States but most widely found in the warm humid and transition regions, and to lesser extent in the cool humid regions.
- Grows well on clay-loam, and sandy soils.
- Medium tolerance to wet and drought soil conditions.

Morphology/Growth Pattern
- It has a prostrate or ascending growth habit with stems that root at the nodes; can grow to 3.0 ft (1.0 m) in height.
- Reseeds prolifically and can be managed to reestablish each year.
- Makes most of its growth in mid-summer and growth slows significantly when temperatures drop to 60°F degrees at night.
- Reproduces by seeds and stolons.

Use and Potential Problems
- It is grazed and harvested as hay.
- It is highly palatable and has high nutritive value.
- Can become invasive because of its thick stands and rapid growth, especially in new spring plantings of other small seeded forages.

Toxicity/Disorders
- None recorded.

Reference
- Crabgrass [9,17,26,32,44,48,55,58,70,72]

Dallisgrass—*Paspalum dilatatum* Poir.

Dallisgrass seeds

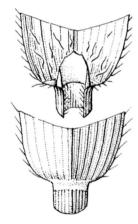

Dallisgrass collar region

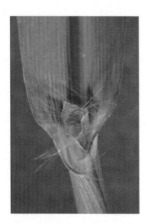

Dallisgrass collar *Source: NCSU*

Spikelets hairy at the base

Seed
- Size: 3—3.5 mm long, 2—2.5 mm wide
- Shape and texture: oval, flat on one side; covered with fine silky hairs looks like a fuzzy tomato seed
- Color: brown or straw-colored
- Seeds per pound:

Dallisgrass seedhead *Source: NCSU*

Leaf sheath and leaf blade—Vernation: Leaf rolled in the bud; **Leaf sheath**: distinctly flattened, lower leaf sheath pubescent (hairy), the upper is glabrous (no hair), split with overlapping hyaline margins; **Leaf blades** are flat, 3—12 mm wide, glabrous, leaf margins with fine hair resembling broadleaf signalgrass (*Brachiaria platyphylla* L.), however, broadleaf signalgrass has much shorter, wider leaves; **Collar**: collar broad with a few long silky hairs; **Auricle**: absent; **Ligule**: long and membranous with a fringe of hairs.

Stolon/Rhizome/Roots: No stolons; *short rhizomes; has fibrous roots.*

Inflorescence: Seedhead produced on a long terminal culm; 3—5 spike-like branches per culm; spikelets arranged on one side of rachis; flattened spikelets, rounded, pointed at the tip; spikelets hairy at the base with a single seed.

Dallisgrass—*Paspalum dilatatum* Poir.

Other common names: Water paspalum, caterpillar grass
Family: Poaceae (Gramineae)
Life cycle: Warm-season perennial
Native to: South America

Distribution and Adaptation
- Best adapted to warm humid and southern parts of the transition zone.
- Widely distributed in pastures, hayland, roadsides, and turf areas.
- It also performs well under short term moisture stress
- Is well adapted to poorly drained soils.
- Rarely found in pure stands and most often found in pastures containing many other warm season species.

Morphology/Growth Pattern
- Relatively long growing season in warm humid zone, but produces most of its growth in June–August.
- If not mowed/grazed can reach 1.6–5.0 ft (0.5–1.5 m) in height.
- Maintains its green condition later into the fall than bermudagrass; appears that frosts does not damage the tissue as much as sorghums and bermudagrass.
- Leaves are arranged close to the soil surface which makes it adaptable to close defoliation.
- Reproduces by seed.

Use and Potential Problems
- Primarily used for grazing, but may be harvested for hay.
- Is a palatable grass, with quality higher than bahiagrass or bermudagrass.
- Appears to maintain nutritive value after early frosts better than bermudagrass.
- Difficult to establish due to poor seed quality, slow germination, and weak seedling vigor.
- Withstands relatively close and frequent defoliation and trampling when well established due to its morphological characteristics (short rhizomes and heavy tillering).

Toxicity/Disorders
- None recorded.

Reference
- Dallisgrass [7,32,55,57,71,72]

Deertongue Grass—*Dichanthelium clandestinum* (L.) Gould (Previously Known as *Panicum clandestinum* L.)

Deertongue seeds

Deertongue collar region

Deertongue plant

Seed
- Size: 1.5 mm long - 2.0 mm wide
- Shape and texture: Elliptical, glossy and smooth, longitudinally stripped
- Seeds per pound: 400,000 https://plants.usda.gov/factsheet/pdf/fs_dicl.pdf

Deertongue plant

Leaf sheath and leaf blade—Vernation: Leaf rolled in the bud-shoot; **Leaf sheath**: most of lower and upper sheath strongly hairy; the middle of the sheath sometimes glabrous; **Leaf blade**: has broad short leaves; well developed, spreading, primary nerves prominent; **Collar**: narrow; **Auricle**: absent; **Ligule**: densely ciliated, 1 mm in length, membranous occasionally may have fringe of hairs.

Stolon/Rhizome/Roots: No stolons; no rhizomes fibrous root system.

Inflorescence: Open panicle; panicle 8−18 cm long; spikelets 2.7−3 mm long.

Deertongue grass—*Dichanthelium clandestinum* (L.) Gould (previously known as *Panicum clandestinum* L.)

Other common names: Deer-tongue grass or deer-tongue panicgrass
Family: Poaceae (Gramineae)
Life cycle: Warm-season annual
Native to: United States

Distribution and Adaptation
- Common throughout the United States, but most prominent in the humid regions.
- Grows in low fertility, acid, loamy, and sandy soils; prefers coarse to medium soil textures.
- Will tolerate sites with a pH as low as 3.8 and high aluminum concentrations.
- Excellent drought tolerance.
- Found in ditches, edges of streams and marshes as well as pastures and shaded areas.

Morphology/Growth Pattern
- Depending on soil and moisture, grows from 1 to 3.0 ft (0.31−0.91 m) tall.
- Bunch type of growth habit.
- Leaves that resemble a "deer's tongue."
- Reproduces by seed.

Use and Potential Problems
- Ground cover for erodible sandy areas such as acid mine spoils, road banks, ditch banks, and gravel pits.
- Considered a favorable wildlife plant because seeds are eaten by many species of birds.
- Considered an undesirable plant in pastures.

Toxicity/Disorders
- None recorded.

Similar Species
- Deertongue may be confused with *common dayflower* (*Commelina communis*) which also has broad and grass-like leaves and a prostrate growth habit. However, deertongue has a distinct ligule and common dayflower has no ligule.

Reference
- Deertongue grass [8,32]

Fescue, Red—*Festuca rubra* L. subsp. *rubra*

Red fescue seeds

Seed
- Size: 3—4 mm long
- Shape and texture: seed narrow with a deep v-shaped cavity, a narrow strip of palea visible, the deeply notched apex folded
- Color: dark straw colored
- Seeds per pound: 600,000

Leaf sheath and leaf blade—Vernation: Leaf folded in the bud; **Leaf sheath**: round, split part away, may be hairy, lower sheath reddish; **Leaf blade**: 0.5—1.5 mm wide, v-shaped, deeply ridged on surface; **Collar**: narrow, continuous, without hairs; **Auricle**: absent; **Ligule**: membranous, long, truncate, 0.2—0.5 mm.

Stolon/Rhizome/Roots: No stolons; slender, short rhizomes.

Inflorescence: Compact panicle.

Fescue, red—*Festuca rubra* L. subsp. *rubra*

Other common names: Creeping fescue, chewing's fescue, flat-leaf red fescue https://plants.usda.gov/plantguide/pdf/pg_feru2.pdf
Family: Poaceae (Gramineae)
Life cycle: Cool season, perennial
Native to: Europe

Distribution and Adaptation
- Introduced from Europe.
- Widely used in the more humid regions of Canada and the northeastern United States.
- Well adapted to well-drained, moderately shaded sites and droughty infertile acid soils; intolerant of wet conditions and high fertility; best adapted to cooler climate and very tolerant of cold temperatures but not of heat and drought.

Morphology/Growth Pattern
- Creeping red fescue is easily distinguished from chewing and hard fescues by its rhizome where chewing and hard fescues are bunch grasses.
- Creeping red fescues has at least three subspecies including ssp. *rubra* and *litoralis*. While both have rhizomes, *rubra* is considered as being a stronger spreading type.
- Reproduces by seed and rhizomes.

Use and Potential Problems
- Mainly used for turf.
- Low mowing heights of 1−2 in. (2.54−5.1 cm)

Toxicity/Disorders
- Alkaloids produced by fungal endophytes.

Reference
- Red fescue [11,12,32,45,64,65]

Fescue, Tall—*Festuca arundinacea* Schreb.

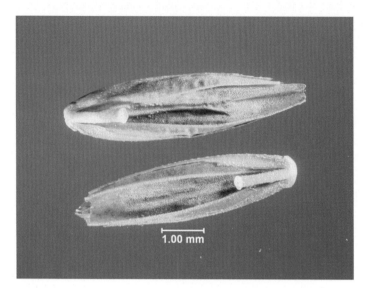

Tall fescue seeds

Seed
- Size: 4–7 mm long, about the same size as perennial ryegrass
- Shape and texture: elliptical and awned rachilla round and knobbed; seed enclosed in lemma and palea
- Color: tan colored
- Seeds per pound: 220,000

Tall fescue plant *Source: NCSU*

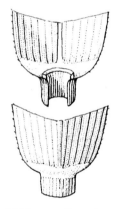

Tall fescue collar region

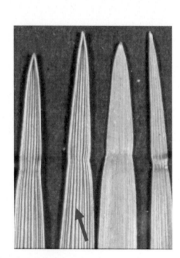

Tall fescue leaf blade

Leaf sheath and leaf blade—Vernation: Leaf rolled in the bud-shoot; **Leaf sheath**: round, smooth, split; **Leaf blades** are flat, broad, glossy on the underside and distinctly barbed and rough on the margins; distinctly veined upper surface, sharp pointed; **Collar**: hairy, broad; **Auricle**: short auricle on most tillers and hairs on margin; **Ligule**: membranous, indistinctly blunt and often not very prominent.

Stolon/Rhizome/Roots: No stolons; shows a few short rhizomes; fibrous roots.

Inflorescence: The inflorescence is a compact panicle, 7.5–10 cm long with lanceolate spikelets; spikelets containing six to eight florets.

Tall fescue seedhead *Source: NCSU*

Fescue, tall—*Festuca arundinacea* Schreb.

Other common names: Reed fescue, coarse fescue
Family: Poaceae (Gramineae)
Life cycle: Cool-season perennial
Native to: Europe

Distribution and Adaptation
- Well adapted to the transition, cool humid, and to a limited extent in the warm humid areas.
- It grows on a wide range of soils, but grows best on loams and clay loams, but does not do well on sandy loam or courser textured soils.
- Grows on soils that vary from strongly acid (pH 4.5) to alkaline (pH 9.5).
- If found on poorly drained soils and will withstand overflow conditions during the cool season, for several weeks.
- It is reasonably tolerant of moisture stress, especially when 2−4 in. of leaf area is remaining on the plant.
- Exhibits good shade tolerance

Morphology/Growth Pattern
- Classified as a bunch-type growth habit but with short rhizomes it forms a dense sod.
- Has weak apical dominance (tillering is not controlled by stem elongation/seed production).
- Reaches a height of 2−3 ft (0.61−91 m).
- Produces one reproductive cycle per year (needs cold temperature and short day for floral induction and warm-temperature and long day for floral initiation).
- Growth pattern extends from late winter through fall with some growth when temperatures are in the 40°F range; does not make much growth when temperatures above 85°F.
- Reproduces by seed and short rhizomes.

Use and Potential Problems
- Widely used for hay, grazing, turf, and to revegetate disturbed sites (mine spoils, roadsides, construction sites, waterways, etc.).
- Most tall fescue naturally infected to varying degrees with a fungal endophyte called *Neotyphodium coenophialum* which is toxic to grazing animals (domestic and wildlife).
- The toxic element is concentrated on the seedhead, stem, and leaf sheath.
- Fescue also is susceptible to nitrate accumulation under conditions of high nitrogen fertilization.
- Is not considered good wildlife habitat, especially for small game, because of the dense canopy and toxic endophyte.
- Reproduces by seed and short rhizomes.

Toxicity/Disorders
- Alkaloids produced by fungal endophytes.

Reference
- Tall fescue [2,4,7,12,29,32,56,57,65]

Foxtail, Giant—*Setaria faberi* **R.A.W. Herrm.**

Giant foxtail seeds

Giant foxtail collar

Giant foxtail seedhead (NCSU)

Seed

- Size: 2.2−2.5 mm long; 1.2−1.5 mm wide
- Shape and texture: ovate and approximately; granular or slightly cross-ridged
- Color: yellowish-green, brown or black depending on seed maturity
- Seeds per pound:

Giant foxtail leaf blade

Leaf blade and leaf sheath—Vernation: Leaf rolled in the leaf bud; **Leaf blade**: giant foxtail may be identified by the presence of many short hairs on the upper surface of the leaf blades, unlike the other foxtails; **Leaf sheath**: round or slightly flattened, split, hairy on margins; **Collar**: broad, may be divided; **Auricle**: absent; **Ligule**: blunt, small, membranous, may have fringe of hairs.

Stolon/rhizome/root system: No stolons; no rhizomes; shallow fibrous roots.

Inflorescence: Dense 7−20 cm long, bending near the base thus the head dropping or nodding.

Foxtail, Green—*Setaria italica* (L.) P. Beauv.: Commonly Referred to as *Setaria viridis* (L.) P. Beauv.

Green foxtail seeds (palea covers more than half of the seed)

Green foxtail collar region

Green foxtail seedhead

Seed

- Size: 1.8−2.2 mm long, 1.0−1.3 mm wide, narrow, and smaller than yellow foxtail
- Shape and texture: spikelet oval to ovate, palea covers more than half of the seed
- Color: seeds are green color
- Seeds per pound:

Green foxtail plant

Leaf blade and leaf sheath—Vernation: Leaf rolled in bud-shoot and has hairs only on the lower margins and leaves with no hairs, which helps to distinguish this plant from giant foxtail and yellow foxtail; **Leaf blade**: 20.3 cm long, 0.42−1.0 cm wide, sharp pointed, mid-vein prominent; **Collar**: broad, divided, hair at least on margins; **Auricle**: absent; **Ligule**: short membrane with fringe of hair.

Stolon/rhizome/root system: No stolons; no rhizomes; fibrous roots.

Inflorescence: The panicle is dense, 2.5−7.5 cm long and covered with many stiff upright bristles; seedhead erect except near tip which is slightly bending.

Foxtail, Yellow—*Setaria pumila* (Poir.) Roem. & Schult. subsp. *pumila* (Formerly Known as *Setaria glauca*)

Yellow foxtail (palea covers less than half of the seed)

Seed

- Size: 2.5–3 mm long, 1.8–2.1 mm wide; larger than green foxtail or barnyardgrass
- Shape and texture: palea side flat or concave; lemma roughened by crosswise wavy ridges; palea covers less than half of the seed
- Color: yellowish green to back depending on maturity
- Seeds per pound:

Yellow foxtail field (NCSU)

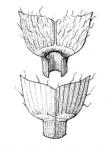

Yellow foxtail collar region

Yellow foxtail collar region

Leaf blade and leaf sheath—Vernation: Leaf rolled in the bud shoot; **Leaf sheath** split, no hairs on the leaf sheath margin as on green foxtail. **Leaf blade**: 4–10 mm wide, 5–30 cm long, flat, sofeet, drooping, taper-pointed, light green; twisted hairs near base, margins smooth or slightly rough. **Collar**: narrow to broad; **Auricle**: absent; **Ligule**: fringe of short hairs 1 mm long or less, fused at base.

Stolon/rhizome/root system: No stolons; have short rhizomes; fibrous roots.

Inflorescence: Dense spikes with yellow to reddish bristles; 5–12 cm long.

Yellow foxtail seedhead (NCSU)

Foxtail, green—*Setaria italica* (L.) P. Beauv.: commonly referred to as *Setaria viridis* (L.) P. Beauv

Other common names: Green bristlegrass, green pigeon grass, wild millet, bottle-grass

Foxtail, giant—*Setaria faberi* (L.) Beauv.

Other common names: Chinese millet, tall foxtail

Foxtail, yellow—*Setaria pumila* (Poir.) Roem. & Schult. subsp. *pumila* (formerly known as *S. glauca*)

Other common names: Summer-grass, golden foxtail, wild millet, bristlegrass, pale pigeon grass
Family: Poaceae (Gramineae)
Life cycle: Warm-season annuals
Native to: Europe-Asia

Distribution and Adaptation
- The three foxtails are distributed throughout the United States.
- The foxtails occur on a wide range of soils but are best adapted to moist, fertile soil during the mid to late summer season.
- Found in new plantings, open turf, or bare spots, open disturbed sites, along forest margins, right-of-ways and old fields.

Morphology/Growth Pattern
- Stem height range from 1 to 6 ft (0.30—1.8 m).
- Giant foxtail is the largest of the three species.
- Yellow foxtail is similar to green foxtail but with long twisted hairs near the base of leaf blades.
- Yellow foxtail branch at the base but does not root at the nodes.
- All three species have a bunch type of growth.
- Reproduces by seed.

Use and Potential Problems
- Yellow and green foxtails most likely found on grazed land.
- The millet-like seed is important food source for wildlife birds.
- Immature foxtail leaves consumed by wild turkeys and may be grazed by domestic livestock.
- Nutritive value in the vegetative stage can be useful for domestic livestock.
- Can accumulate nitrates.

Toxicity/Disorders
- Not a toxic but mechanical injury from awns.

Similar Species
- The three foxtails *green foxtail, giant foxtail, and yellow foxtail* resemble each other in that all the three foxtails have a dense cylindrical clustered seedhead. Yellow foxtail seedhead has yellow bristle, the leaf sheath is smooth and the leaf blade is hairy only at the base close to the collar region; green foxtail has green bristles, the upper leaf surface is smooth and hairs are on the leaf sheath while giant foxtail has a nodding seedhead with green bristles, and the leaf blade is hairy on the entire upper surface, the leaf sheath has fine hairs on some part of the margin.

Reference
- Foxtails [17,18,26,32,34,36,45,46,56,57,65,71,75]

Gamagrass, Eastern—*Tripsacum dactyloides* L.

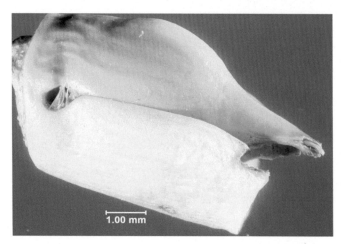

Gamagrass seed

Gamagrass collar region

Eastern gamagrass stem and short rhizomes

Eastern gamagrass seedhead

Seed
- Size: 4 mm long
- Shape and texture: broadly conical
- Color: straw to reddish
- Seeds per pound: 7400

Eastern gamagrass plant

Leaf sheath and leaf blade—Vernation: Leaf rolled in the bud-shoot; **Leaf sheath**: flattened, smooth, split with overlapping margins; **Leaf blade**: 20–25 mm wide, rough above and below at least near tip; **Collar**: narrow to broad, sometimes divided; **Auricle**: absent; **Ligule**: a fringe of hairs, sometimes fused at base.

Stolon/Rhizome/Roots: No stolons; short rhizomes; fibrous roots.

Inflorescence: With one to three terminal inflorescence; has male spikelets above and female spikelets below; the male spikelets are in pairs fitting into the rachis; the female spikelets are oval and hard having woody joints.

Gamagrass, eastern—*Tripsacum dactyloides* L.

Other common names: Wild Corn, "Ice Cream Grass"
Family: Poaceae (Gramineae)
Life cycle: Warm-season perennial
Native to: United States

Distribution and Adaptation
- Is found widely over the United States, especially east of the Central Plains. Adapted to fine-, medium-, and coarse-textured soils.
- Tolerates poorly drained soils, but does not tolerate extended inundation.
- Grows well at pH range of 5.5−7.5.
- It is a distant relative of corn.
- It is relatively tolerant of long-term moisture stress that is encountered in the humid regions of the country.

Morphology/Growth Pattern
- Leaf mass may reach 2−3 ft height (0.61−0.91 m), but flowering stems may reach 7 ft (2.1 m) or more. The inflorescence has both male and female parts in the same spike (male spikelets above and female spikelets below). The plant forms large clumps or crowns at maturity and the center of the clump tends to "die-back" and enlarge (up to 3′ across) with age (see photo).
- Most of the spreading comes from thick, knotty, short-jointed rhizomes.
- Plants produce few seeds and dormancy is high.
- Mature stands have significant bare soil between the clumps, but large leaves develop forming complete canopy cover when vegetation reaches 12−20 in. (30.5−50.8 cm) height. It makes most of its annual growth in June−August but has it starts new spring growth early and continues until frost.
- Reproduces by seed and short rhizomes.

Use and Potential Problems
- Primarily used for grazing, but also used for hay, silage, erosion control, and wild life habitat.
- Major challenges occur during establishment because of the high dormancy of seeds and the relatively slow seedling development.

Toxicity/Disorders
- None recorded.

Reference
- Eastern gamagrass [4,7,29−32,56−58]

Goosegrass—*Eleusine indica* (L.) Gaertn.

Goosegrass seeds

Seed
- Size: 1–2 mm long
- Shape and texture: each spikelet contains three to six seeds; granular and ridged
- Color: light brown to black
- Seeds per pound:

Goosegrass plant (NCSU) Goosegrass field

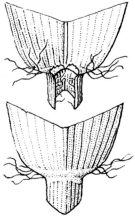

Collar region of goosegrass

Goosegrass collar region

Leaf sheath and leaf blade—Vernation: Leaf folded in the bud-shoot; **Leaf sheath**: distinctly flattened, hairy at top, split with overlapping margins; sheath near the base is silver or white; **Leaf blade**: glabrous, or occasionally sparsely hairy; **Collar**: broad, hairy on margins; **Auricle**: absent; **Ligule**: membranous, 0.6–1 mm long, truncate, sometime short ciliate.

Stolon/Rhizome/Roots: No stolons; no rhizomes; fibrous roots.

Inflorescence: Spikelets crowned on one side of rachis in two rows, spikelet branches do not all originate from same point; spikelets are longer than bermudagrass, crabgrass, and Rhodes grass.

Spikelet branches not all originate from the same point http://www.psu.missouri.edu/fishel/goosegrass.htm

Goosegrass—*Eleusine indica* (L.) Gaertn.

Other common names: Wiregrass, yardgrass, silver crabgrass, crowfootgrass, Indian eleusine, bullgrass
Family: Poaceae (Gramineae)
Life cycle: Summer annual
Native to: Europe

Distribution and Adaptation
- Found throughout the United States, with the exception of the arid regions in western states.
- Found in full sun and in poorly drained or compacted soil.
- Mostly found in pastures, disturbed areas, or uncultivated sites.

Morphology/Growth Pattern
- Produces a prostrate, mat-like basal rosette, with stems spreading from the central point.
- Grows 0.5−2 ft (0.15−0.61 m) high
- Major differences between goosegrass and crabgrass are:
  - Goosegrass has flat stems and does not root at the lower nodes like crabgrass.
  - Stems branched and arise from tufts.
  - Goosegrass is darker green than crabgrass.
  - It tolerates close mowing, compacted soils, and drought.
- Reproduces by seed.

Use and Potential Problems
- Very undesirable plant in forage systems because of its high tensile strength and unpalatability after stem elongation begins.
- Primarily a weed of compacted pasture areas (around water tanks, lounging areas, and trails), nursery, and agronomic crops.
- Is a very troublesome weed in athletic fields, golf greens, and fairways.

Toxicity/Disorders
- None recorded.

Reference
- Goosegrass [17,24,32,33,44,48,57,65,71,75]

Indiangrass—*Schizachyrium scoparium* L.

Indiangrass seed

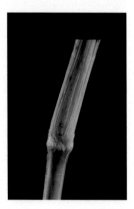

Indiangrass node with hairs

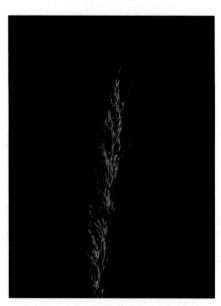

Indiangrass seedhead

Seed
- Size: 3—4 mm long
- Shape and texture: Obovate with long tapering to a long pointed base https://plants.usda.gov/factsheet/pdf/fs_sonu2.pdf
- Color:
- Seeds per pound: 192,000

Indiangrass leaves and hairy stem

Leaf sheath and leaf blade—Vernation: Leaf rolled in bud-shoot; **Leaf sheath**: generally shorter than internodes; **Leaf blade**: 25.4—61 cm long, flat, tapered to a narrow base where it joins the sheath, sometimes hairy, light green; **Collar**: broad; **Auricle**: absent; **Ligule**: membranous, claw-like, 1.3 cm long, notched at tip.

Stolon/Rhizome/Roots: No stolons; short rhizomes; fibrous roots.

Inflorescence: Long, dense, golden panicle; 10.2—20.4 cm long, spikelets paired hairy (fuzzy).

Indiangrass—*Schizachyrium scoparium* L.

Other common names: Yellow Indiangrass, prairie plume grass
Family: Poaceae (Gramineae)
Life cycle: Warm-season, perennial
Native to: United States

Distribution and Adaptation
- Found east coast to the Rocky Mountains and Arizona.
- Best adapted to deep moist soils from heavy clay to deep sands.
- Found on roadsides and areas where little mowing or grazing are routinely practiced.
- Tolerates sites that are slightly acidic or alkaline.

Morphology/Growth Pattern
- Bunch-type growth habit.
- Grows up to7 ft (2.13 m).
- Begins growth in spring later than switchgrass, making most of its growth between June and August, but remains green till frost.

Use and Potential Problems
- Used primarily for grazing, but may be harvested for hay.
- Highly palatable and of high nutritive value when harvested before seed set.
- Due to its attractive seedhead at maturity, the plant is used in gardens and flower arrangements.
- Persists best when defoliated infrequently and not below 6 in. (15.24 cm).
- Planting the "fluffy" seeds can be a challenge and seedling vigor is relatively weak.
- Reproduces by seeds and short rhizomes.

Toxicity/Disorders
- None recorded.

Reference
- Indiangrass [7,28,32,34,39,54,55,57,58]

Johnsongrass—*Sorghum halepense* (L.) Pers.

Johnsongrass seeds

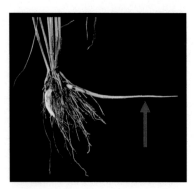

Johnsongrass rhizomes

Johnsongrass seedhead

Seed

- Size: 2—3 mm long, 1.3—1.8 mm wide
- Shape and texture: look for knobbed rachilla on seed; surface ridged longitudinally
- Color: dark black to straw colored
- Seeds per pound:

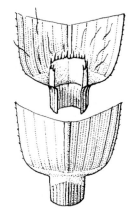

Collar region of Johnsongrass

Johnsongrass plant (with Elizabeth Yarber)

Leaf sheath and leaf blade—Vernation: Leaf rolled in the bud-shoot; **Leaf sheath**: round to somewhat flattened, smooth, split with overlapping margins; **Leaf blade**: usually without hairs (glabrous) on both surfaces, however, some hairs may be present at the base of the leaf blade. Leaves have a prominent whitish mid-vein, which snaps readily when folded; **Collar**: broad, sometimes divided, may be hairy; **Auricle**: absent; **Ligule**: membranous, 3—6 mm long, rounded to acute and occasionally toothed, consists of a fringe of dense, fine hairs.

Stolon/Rhizome/Roots: No stolons: many thick rhizomes; fibrous roots.

Inflorescence: Large loose, erect open panicle; well-branched, and often has a reddish tinge to it.

Johnsongrass—*Sorghum halepense* (L.) Pers.

Other common names: Mean-grass, Aleppo grass, grass sorghum, Egyptian millet
Family: Poaceae (Gramineae)
Life cycle: Warm-season perennial
Native to: Mediterranean region

Distribution and Adaptation
- Found throughout warm humid and transition zone and to lesser extent in the northern parts of the cool humid region.
- Found along roadsides, waste areas and fields, fencerows, and ditch banks, preferably in moist sites in cultivated fields (row crops and hay) and pastures.
- Grows best on heavier soils but performs moderately well on fine sandy loams.
- Responds well to soil fertility.

Morphology/Growth Pattern
- It spreads by seed in the north, but in central and southern United States it also spreads by rhizomes.
- Depending on soil type, plant may reach 6−7 ft (1.8−2.13 m).
- Grows after soil temperatures warm until frost.
- Makes most of its growth in June−August.
- Reproduces by seed and rhizomes.

Use and Potential Problems
- Used for hay, grazing, and erosion control, provides good cover, and seed provides excellent food source for many species of birds.
- Originally was brought to the United States for livestock feed, but now it is ranked as noxious weed.
- When the plant is stressed (wilting caused by drought, frost, herbicide spray), or mechanically injured (trampled), a great deal of free hydrocyanic acid (prussic acid) is produced; upon ingestion, the animals quickly develop signs related to cyanide poisoning.
- Johnsongrass, like other plant species, can accumulate toxic levels of nitrates, depending on fertilization practices.

Toxicity/Disorders
- Nitrate, prussic acid.

Similar Species
- *Shattercane (Sorghum allum* L.) is closely related to Johnsongrass; morphologically, it is difficult to distinguish the two plants; however, shattercane is an annual plant and does not have rhizomes. Johnsongrass also can be mistaken for *sudangrass (Sorghum vulgare sudanense* L.), however, sudangrass is an annual without rhizomes. Additionally, the seeds of sudangrass are brown in color, while those of Johnsongrass are reddish-brown in color.

Reference
- Johnsongrass [1,2,14,17,26,32,33,44,48,49,56,57,65,67,71,72,74−76]

Lacegrass—*Eragrostis capillaris* (L.) Nees

Lacegrass seeds

Lacegrass collar region

Seed
- Size: grain 0.6 mm long; lemma and palea 1.2 mm long
- Shape and texture: oblong, grooved
- Color: reddish
- Seeds per pound:

Lacegrass collar region

Leaf sheath and leaf blade—Vernation: rolled in bud-shoot; **Leaf sheath**: round, smooth, split with overlapping hyaline margins; **Leaf blade**: 2−3 mm wide, rough above and below and on margin; **Collar**: broad; **Auricle**: absent; **Ligule**: membranous, 1−2 mm long, rounded, toothed.

Stolon/Rhizome/Roots: No stolons; no rhizomes; fibrous roots.

Inflorescence: Panicle very long and wide, covers two-thirds of the entire length of the plant

Lacegrass—*Eragrostis capillaris* (L.) Nees

Other common names: English sundew, tiny lovegrass, lace lovegrass
Family: Poaceae (Gramineae)
Life cycle: Warm-season, annual
Native to: Europe

Distribution and Adaptation
- Found in much of the region east of the Plains states.
- Found in open, disturbed, or cultivated soil.
- Usually found on well-drained soils.

Morphology/Growth Pattern
- Tall, freely branching at the base.
- Tufted, often ascending. Bunch type of growth habit.
- Is probably not very competitive with other species.
- Reproduces by seed.

Use and Potential Problems
- It is an undesirable plant in forage systems and has little wildlife habitat value.
- Nutritive value in the early vegetative stage is probably good.

Toxicity/Disorders
- None recorded.

Reference
- Lacegrass [8,32,56,57]

Lovegrass—*Eragrostis curvula* L.

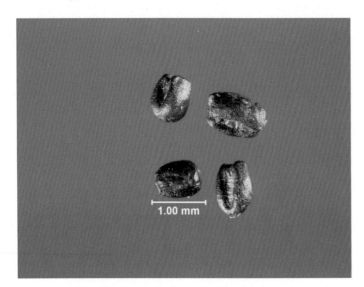

Lovegrass seeds

Collar region of weeping love grass

Weeping lovegrass seedhead

Seed
- Size: 1.75–2 mm long, 1 mm wide
- Shape and texture: oval, embryo arca oblong with a distinct rim; smooth or faintly striate
- Color: reddish purple to black in color germ has a lighter color
- Seeds per pound: 750,000

Weeping lovegrass plant

Leaf sheath and leaf blade—Vernation: Leaf rolled in bud-shoot; **Leaf sheath**: round, smooth, split with overlapping hyaline margins; **Leaf blade**: 2–3 mm wide, rough above and below and on margin; fine textured, almost hair-like leaf blade; **Collar**: broad; **Auricle**: absent; **Ligule**: membranous, 1–2 mm long, rounded, toothed.

Stolon/Rhizome/Roots: No stolons or rhizomes; bunch type of growth habit.

Inflorescence: Panicle 20.0–30.5 cm long and somewhat drooping; flowers are olive-purple, maturing to a grayish color, and are quite delicate.

Lovegrass—*Eragrostis curvula* L.

Other common names: Ticklegrass, African lovegrass
Family: Gramineae/Poaceae
Life cycle: Warm-season perennial
Native to: South Africa

Distribution and Adaptation

- Weeping lovegrass is best adapted to warm humid and transition zones, but is found on roadsides in Pennsylvania and New Jersey.
- One native and three introduced species. Sand lovegrass is native to the central and southern Great Plains.
- Very drought tolerant; grows well in hot, dry locations in full sun and in most soils if well drained; well adapted to sandy soils.

Morphology/Growth Pattern

- A vigorous-growing perennial bunchgrass.
- When moisture is adequate, plants reach a height of 2−4 ft (0.60−1.2 m).
- Produces numerous slender basal leaves, 0.8−1.66 ft (0.25−0.50 m) long and 1/16 in. (0.16 cm) wide that taper to a needle like point. Bases of the densely clustered young leaves are often purplish; the leaf blades as a whole are light green.
- Easily established by seeds and spreads only a few inches by tillering.
- Canopy often thick and drooping or lodged.
- Reproduces by seed.

Use and Potential Problems

- As a forage, it is used primarily for grazing, mainly in the southern Plains.
- Used extensively for quick soil cover on disturbed sites for temporary erosion control, however, as stands mature the clumpy nature of the stand is subject to soil erosion. Commonly planted on highway embankments during summer months.
- Used extensively as an ornamental grass.
- If not cut, stands of high mass are a potential fire hazard during the winter dormant period.
- Clumpy stands may provide habitat for wildlife.

Toxicity

- None recorded.

Reference

- Lovegrass [7,27,48,49,56,57,73,75]

Millet, Foxtail—*Setaria italica* (L.) P. Beauv.

Foxtail millet seeds

Foxtail millet collar region

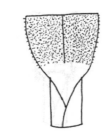

Japanese millet collar region

Foxtail millet plant *Source: Chris Teutsch*

Seed
- Size: 2–2.25 mm long, 1–1.5 mm wide
- Shape and texture: foxtail millet seeds flat on one side, horseshoe on germ side; ridge runs lengthwise down middle of seed; lemma and palea usually attached to seed; entire surface of lemma tubercled or along the margins
- Color: golden yellow
- Seeds per pound: 220,000

Foxtail millet plant *Source: Rocky Lemus*

Foxtail millet

Leaf sheath and leaf blade—Vernation: Leaf rolled in bud-shoot; **Leaf sheath**: round with sparse short hairs, split with overlapping hairy margins; **Leaf blade**: wide, rough above and below at least near tip, margins very rough; **Collar**: medium broad to broad, hairy, also hair on margins; **Auricle**: absent; **Ligule**: a fringe of hairs and sometimes fused at the base

Stolon/Rhizome/Roots: No stolons or rhizomes; bunch type of growth habit with fibrous root systems.

Inflorescence: Spike-like, dense, bristly, cylindrical, large yellow bristles resembling yellow foxtail, green foxtail, or giant foxtail. (Japanese millet: compact panicle with short, densely flowered, spike-like branches.)

Japanese millet

Leaf sheath and leaf blade—Vernation: Leaf rolled in bud-shoot; **Leaf sheath**: round to somewhat flattened, smooth, split with overlapping margin; **Leaf blade**: wide, rough above and below at least near tip, margins very rough; **Collar**: narrow; **Auricle**: absent; **Ligule**: absent.

Millet, foxtail—*Setaria italica* (L.) P. Beauv.

Other common names: German or Italian millet
Family: Gramineae/Poaceae
Life cycle: Warm-season, summer annuals
Native to: Southern Asia

Distribution and Adaptation
- Grown in the warm humid and transition zones primarily, but also in the cool humid zone.
- Compared to other cereal grains, generally grown on less fertile soils in hot and low moisture environments.
- Grown on a wide range of soil types, best suited to well-drained sandy soils.
- Less tolerant to poorly drained soils and flooding than sorghum or sorghum-sudangrass hybrids.

Morphology/Growth Pattern
- Robust, quick growing, with large stems and leaves.
- Erect and 2−4 ft (0.60−1.2 m) tall with a panicle inflorescence made up of 5−15 sessile erect branches. Panicle, large, heavy, dense, cylindrical to lobed, sometimes interrupted at base, erect or often nodding.
- Generally do not make significant regrowth when cut or grazed after stem elongation.
- Reproduces by seed.

Use and Potential Problems
- Less productive than pearl millet or sorghum-sudan hybrids; used mainly for hay.
- Also useful as a cover crop and for short season production, especially as emergency pasture or hay and often grown in mixtures with soybeans or cowpeas.
- Often planted on disturbed sites to provide quick soil cover to minimize erosion.
- Food for songbirds and game birds.

Toxicity
- Photosensitization.

Reference
- Millets [4,12,31,32,49,58]

Millet, Japanese—*Echinochloa frumentacea* Link

Other common names: Barnyard millet or billion dollar grass
Family: Gramineae/Poaceae
Life cycle: Warm-season, summer annual
Native to: Southern Asia

Distribution and Adaptation
- Will grow better in cool regions than foxtail millet; in the United States. Mostly grows in New England and along the upper Atlantic Coast area.
- Compared to other cereal grains, generally grown on less fertile soils in hot and high moisture environments.
- Grown on a wide range of soil types, but are best suited to sandy soils.
- Japanese millet is more tolerant to poorly drained soils and flooding than sorghum or sorghum-sudangrass hybrids.

Morphology/Growth Pattern
- Taller, coarser growing plant than any of the foxtail millets; robust, quick growing, with large stems and leaves.
- Growth habit is erect and 2—4 ft (0.60—1.2 m) tall.
- Generally do not make significant regrowth when cut or grazed after stem elongation.
- Reproduces by seed.

Use and Potential Problems
- Mostly used for hay, pasture, and silage.
- Useful for a cover crop, wildlife feed, and erosion control on wet natured soils.
- Food for songbirds and game birds.

Toxicity
- Photosensitization.

Reference
- Millets [4,12,31,32,49,58,67,75]

Millet, Pearl—*Pennisetum americanum* (L.) Nash

Pearl millet seeds

Pearl millet collar region

Pearl millet hairy collar region and leaf sheath

Seed
- Size: 3—3.5 mm long, 2—2.5 mm wide
- Seed shape and color: obovate, and thickened toward the top, in side view, the dorsal side is straight, the ventral side is arched; smooth, fairly shiny
- Color: white to gray
- Seeds per pound: 86,000—88,000

Pearl millet plant

Leaf sheath and leaf blade—Vernation: Leaf rolled in the bud; **Leaf sheath**: hairs on the margins of the leaf sheath; **Leaf blade**: wide; **Collar**: medium broad to broad, divided, hairy; **Auricle**: absent; **Ligule**: membranous.

Stolon/Rhizome/Roots: No stolons or rhizomes, fibrous root system.

Inflorescence: The numerous flowers are tightly tucked around a compact cylindrical, spike-like seedhead, the length of the seedhead ranges from 15 to 40 cm; hair under seedhead.

Millet, pearl—*Pennisetum americanum* (L.) Nash

Synonyms: *Pennisetum typhoides* (Burm) Stapf. & Hubb., *P. typhoideum* (L.) Rich., and *P. americanum* (L.) Leeke
Other common names: Cattail millet
Family: Gramineae/Poaceae
Life cycle: Warm-season, summer annual
Native to: West Africa

Distribution and Adaptation
- Adapted to most medium- or well-drained soils except sandy soils. Better performance on sandy loams of the coastal plain soils than sorghum-sudan hybrids.

Morphology/Growth Pattern
- Plant erect, more leafy than sorghum-sudan hybrids.
- Grows 3—6 ft (0.9—1.8 m) tall.
- Flowers can be either cross-pollinated or self-pollinated.
- Pearl millet can be classified into three categories based on height: dwarf less than 4 ft (1.2 m), semidwarf 4—6 ft (1.2—1.8 m, and tall 6—8 ft (1.8—2.4 m). The taller type has more stem and produces more dry matter.
- Reproduces by seed.

Use and Potential Problems
- Used for grazing hay or silage; will produce 3—4 tons of forage per acre on a dry matter basis.
- Difficult to make hay due to its thick stem (requires long drying time unless stem is mechanically crushed).
- Reproduces by seed.

Toxicity
- One of the forage species that accumulates nitrate-nitrogen when applied in a large quantity. Nitrate poisoning in livestock is caused by the consumption of pasture or hay containing high levels of nitrate-nitrogen. Prussic acid which can build up to toxic levels in Johnsongrass, sorghum, sudangrass, sorghum-sudan hybrids is not produced in pearl millet.

Reference
- Millets [4,12,31,32,49,58,67]

Nimblewill—*Muhlenbergia schreberi* J.F. Gmel.

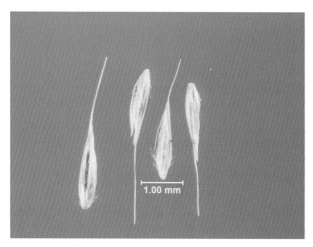

Nimblewill seeds

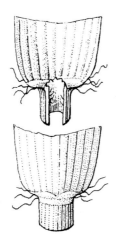

Nimblewill collar region

Nimblewill collar region

Seed
- Size: 1.8−2.2 mm long, 0.4−0.6 mm wide
- Shape and texture: narrowly elliptical, paleas bulged out, straight awn as long or longer than seed, base covered with a few white hairs
- Color: caryopses dark brown and striate; seed coat yellowish or silver

Nimblewill plant *Source: Nathan O'Berry*

Leaf sheath and leaf blade—Vernation: Leaf rolled in the bud-shoot; **Leaf sheath**: flattened, smooth, split, shorter than internode, usually a few long hairs on upper edge and near throat, prominent white membrane along edge; **Leaf blade**: flat, not hairy except for occasional hairs on edge near the base; **Collar**: broad, hairy on edges; **Auricle**: absent; **Ligule**: membranous, minute, irregular edge, jagged, less than 0.5 mm long.

Stolon/Rhizome/Roots: Stolons present, no rhizomes; stems are slender, wiry, round, and smooth; has a fine, fibrous root system; shallow-rooted.

Inflorescence: Loose, spike-like panicle, spikelets have long awns; tiny inconspicuous flowers and seeds on upright branches.

Nimblewill—*Muhlenbergia schreberi* J.F. Gmel.

Family: Gramineae/Poaceae
Other common names: Drop-seed, wire-grass, satin-grass.
Life cycle: Warm-season perennial
Native to: United States

Distribution and Adaptation
- Common throughout the eastern states and appears to be invasive in the transition zone.
- Grows best on shaded sites with moist soils.

Morphology/Growth Pattern
- Is a fine textured, mat-forming blue-green perennial, forming patches in lawns that resemble bentgrass.
- Can grow to 1—3 ft (0.30—0.90 m) long.
- Slender stems and erect early season growth, with short, narrow leaves that spread out horizontally in different directions; stems have knots at the node.
- Early season stems are erect, but sag toward the ground and become spreading with age. Mature stems can root at the nodes and are branched.
- Reproduces by seed and stolons.

Use and Potential Problems
- Troublesome in pastures because it is not readily eaten by grazing animals. However, animals will "learn" to eat it if exposed to the young plants prior to stem elongation.
- May invade lawns, orchards, nurseries, and gardens, and is also commonly found in waste places and roadsides.
- Nutritive value is similar to bermudagrass, but animal preference for it is low compared to bermudagrass.

Toxicity
- None recorded.

Reference
- Nimblewill [33,46,48,49,66,67,72,76]

Nutsedge, Yellow—*Cyperus esculentus* L.; Nutsedge, Purple—*Cyperus rotundus* L.

Yellow nutsedge seeds

Yellow nutsedge tuber

Long linear three-ranked leaves grow end of stem *Source: Peter Sforza*

Seed (Yellow Nutsedge)
- Size: 1.1−1.5 mm long, 0.5−0.7 mm wide
- Shape and texture: ovate, three-sided, broader at the apex, and gradually tapered to narrow, round base, angles blunt, rounded, and the sides distinctly concave; surface of the seed is shiny
- Color: light brown to yellowish-brown

Yellow left and purple nutsedge right plants *Source: NCSU*

Leaf sheath and leaf blade—Leaf: Long, linear, three-ranked leaves may extend above the stem; grass-like, 20−90 cm long and 4−9 mm wide; waxy and glossy with prominent mid-rib; leaf tip sharply pointed; **Leaf sheath**: closed and triangular; **Auricle**: absent; **Ligule**: absent.

Stolon/Rhizome/Roots: No stolons; underground rhizome and tubers; stem triangular, slender, smooth longer or shorter than basal leaves.

Inflorescence: Seedhead produced on a spike in a compound umbel with several leaf-like long bracts beneath flowers often longer than the flower cluster; flower head bristle, small yellow flowers born at the end of the stem similar to a loose cylindrical brush; each ray of the umbel is branched.

Nutsedge, yellow—*Cyperus esculentus* L.

Other common names: Nutsedge, chufa nutgrass, yellow nutgrass, earth almond, nut rush, yellow galingale, northern nutgrass, coco sedge

Nutsedge, purple—*Cyperus rotundus* L.

Other common names: Purple nut-grass, nutgrass, coco-sedge, coco-grass
Family: Cyperaceae
Life cycle: Warm season, perennial
Native to: United States

Distribution and Adaptation
- Found throughout the three zones.
- Often infests wet areas and spreads to other areas.

Morphology/Growth Pattern
- Both species are grass-like plants growing from 1 to 3 ft (0.30–0.90 m) long; shiny yellow-green, leaves which are three-ranked, triangular stem, no leaves on stem.
- Tubers are the only part of the plant that lives through the winter; in a single growing season one tuber can produce 1900 plants and 7000 new tubers in a 2 m^2 area.
- Reproduces by seeds and underground tubers.

Use and Potential Problems
- Domestic herbivores do not readily graze these plants.
- Often a serious weed in pastures, hay and cultivated fields, row crops, horticultural crops, nursery crops, turfgrass, landscape, and forest seedlings.
- Tuber slightly sweet to taste.
- Chufa is a variety or selection of yellow nutsedge; underground tuber has been marketed as a food source for turkeys, ducks, and wild pig.

Similar Species
- *Cyperus* includes some 600 species but the three most common species of nutsedge are: *purple nutsedge* (*C. rotundus* L.), *yellow nutsedge* (*C. esculentus* L.), and *false nutsedge* (*Cyperus strigosus* L.). Purple nutsedge has dull reddish-purple flower cluster, while yellow nutsedge and false nutsedge have yellow to yellowish-green color; yellow nutsedge has small scaly whitish rhizomes ending with a small tuber, lacking hairs; purple nutsedge develops tube-chains along the length of the rhizomes, while false nutsedge has hard perennial base but no rhizome or tubers.

Toxicity
- None recorded.

Reference
- Nutsedge [24,33,46,48,49,66,67,72,76]

Oat, Cereal—*Avena sativa* L.

Oat seed

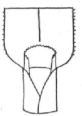

Oat collar region

Oat inflorescence *Source: Stephen Harrison*

Seed
- Size: grain 8–9 mm long; palea 9–15 mm long
- Seed shape and texture: seed remains in the husk (lemma and palea); cultivated oats lack the circular scar (sucker mouth) on the seed and either lack an awn or the awn is straight; lemma smooth
- Color: straw/tan
- Seeds per pound: 16,000

Oat plant *Source: Stephen Harrison*

Leaf sheath and leaf blade—Vernation: Leaf rolled in the bud-shoot; **Leaf sheath**: round, smooth, split with overlapping margins; **Leaf blade**: 10–14 mm wide, smooth near base and rough near tip above and below, margins rough and hyaline; **Collar**: broad, divided; **Auricle**: absent; **Ligule**: membranous, 3–4 mm long, rounded, toothed.

Stolon/Rhizome/Roots: No stolon or rhizome, fibrous root system.

Inflorescence: Loose open panicle.

Oat, cereal—*Avena sativa* L.

Other common names: None recorded
Family: Gramineae/Poaceae
Life cycle: Cool season, annual
Native to: Europe, Iraq, and Turkey

Distribution and Adaptation
- Oat for grain is grown mostly in the cool humid zone, but for winter pasture and forage most is grown in the warm humid and the southern section of the transition zones.
- Grows well on well-drained loams and silt loam soils.
- In the transition zone forage oat generally does not provide as much winter growth as rye, wheat, or triticale, however, in the warm humid region, it performs similarly.

Morphology/Growth Pattern
- Bunch type of growth habit, but tillers sufficiently to fill in between drill rows.
- Depending on nitrogen fertilization and variety, can grow 3 ft (0.90 m) tall or more.
- Winter oat requires period of cold temperature for floral initiation; spring oats do not need cold to initiate flowering; spring oats are not winter hardy and may be killed when temperatures drop to 26−28°F.
- Reproduces by seed.

Use and Potential Problems
- Used for hay, silage, grazing, and grain.
- Used as a companion crop when establishing some forage legumes and grasses, especially in the cool humid zone.
- Excellent rotational crop for wheat or barley because it is not susceptible to the same species of diseases.

Toxicity
- Nitrate, nitrate toxicity, nitrate poisoning.

Reference
- Oat [4,7,12,32,49,58,73]

Oat Wild—*Avena fatua* L.

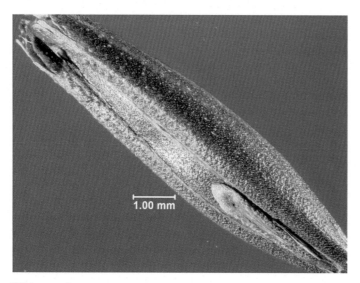

Wild oat seed

Seed

- Size: 11—14 mm long, 2—5.5 mm wide
- Shape and texture: ventral side flat, with the fine groove down the middle; veined glumes; numerous brown hairs at the base of the seed; large rachilla; when present, awn twisted and bent at 90-degree angle
- Color: dark to light brown

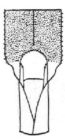

Wild oats collar region

Leaf sheath and leaf blade—Vernation: Leaf rolled in the bud-shoot; **Leaf sheath**: round, smooth, split with overlapping hyaline margins; **Leaf blade**: 10—20 mm wide, rough above, smooth near base and rough near tip, mid-rib prominent, margins rough and in young plants sometimes hairy near base; seedlings twist counter-clockwise when viewed from above; **Collar**: broad, may be divided; **Auricle**: absent; **Ligule**: membranous, 3—5 mm long, rounded, mostly toothed.

Stolon/Rhizome/Roots: No stolons or rhizomes; fibrous root system.

Inflorescence: A loose open panicle; branching and more spreading than the cultivated oat; two or three florets in each spikelet; lemma of each floret is hairy; awns are twisted and bent at right angles when mature.

Oat, wild—*Avena fatua* L.

Other common names: Oat grass, poor oats, wheat oats, flax-grass black oats
Family: Gramineae/Poaceae
Life cycle: Cool-season annual
Native to: Europe and Asia

Distribution and Adaptation
- Found primarily in the transition and humid cool zone.

Morphology/Growth Pattern
- Annual tufted plant with an extensive root system; stem with dark colored nodes; the stem of the mature plant is smooth, erect, and grows up to 4 ft (1.2 m) leaves are similar to cultivated oats.
- In the vegetative stages, sometimes confused with wheat and barley; wheat and barley have clasping auricle wrapped around the stem (very prominent and clasping on barley and thin and hairy on wheat), while wild oats do not have auricles; wild oat stalks are usually much taller than cultivated oat stalks; also the seedheads of wild oats are more open and spreading; when mature are more lopsided and drooping than cultivated oats.
- Reproduces by seed.

Use and Potential Problems
- May be grazed in vegetative stage and would be of high nutritive value.
- Competitive weed during establishment of some crops.
- Causes yield losses, dockage losses, cleaning costs, and lowers grade and quality when mixed with small grains.

Toxicity
- Not a toxin, but mechanical injury from awns.

Reference
- Wild oat [4,12,31,32,58,61,73,75]

Orchardgrass—*Dactylis glomerata* L.

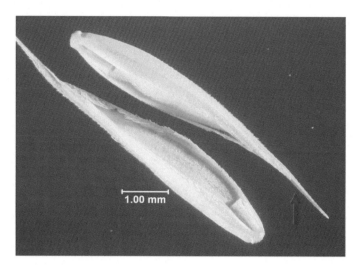

Orchardgrass seeds with twisted lemma end

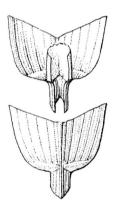

Orchardgrass collar region

Orchardgrass seedhaead *Source: NCSU*

Seed
- Size: grain 3 mm long; lemma 5−7 mm long
- Shape and texture: lemma compressed laterally, the apex commonly bent off-center; caryopsis is soft and shorter than the palea.
- Color: light straw-colored
- Seeds per pound: 590,000 (unhulled), 625,000 (hulled)

Orchardgrass plant *Source: Chris Teutsch*

Leaf sheath and leaf blade—Vernation: Leaf folded in the bud-shoot; **Leaf sheath**: leaves strongly compressed, flattened, smooth to somewhat rough or rarely hairy, mostly split part way only with hyaline margins; **Leaf blade**: 6−12 mm wide, broadly linear, mostly smooth near base and rough near tip; leaf margins rough; **Collar**: broad, divided; **Auricle**: absent; **Ligule**: membranous, 3−10 mm long, mostly acuminate.

Stolons/Rhizomes/Roots: No stolon or rhizome; fibrous root system.

Inflorescence: Panicle with spikelets crowded in dense clusters at end of the few stiff branches; long first panicle branch (short first panicle branch for reed canarygrass); spikelets on one side of rachis, hairy on keels with three to five seeds; seeds have a very short awn; florets are attached by short, thick rachillas.

Orchardgrass—*Dactylis glomerata L.*

Other common names: Cocksfoot
Family: Gramineae/Poaceae
Life cycle: Cool season perennial
Native to: Europe, North Africa, and parts of Asia

Distribution and Adaptation

- Grown primarily in the cool humid and the transition zones. Limited adaptation in the northern areas of the warm humid zone.
- Persists only about 3 years at lower elevations in the transition zone.
- Only moderately winter hardy and frequently will not survive northern climatic conditions if snow cover is lacking.
- Less tolerant of drought and poor drainage than tall fescue.
- Relatively shade tolerant.
- More than 60% of production in spring and fall, and growth is slowed significantly when temperatures exceed 85°F.

Morphology/Growth Pattern

- Bunch grass, tall, upright stem growth; grow 2–3 ft (0.60–0.90 m) tall, fast-growing and matures very early in the spring.
- Carbohydrates are stored in thick, fleshy stem base.
- In general, orchardgrass matures about 1 week earlier than tall fescue and about 2 weeks before smooth bromegrass in the mid-Atlantic region.
- Flowering tillers are initiated by cool short days, therefore they are present only in spring; all subsequent regrowth for the year is leaf.
- During flowering and seed formation plants show strong apical dominance (hormonal control).
- Reproduces by seed.

Use and Potential Problems

- Commonly used for grazing, hay, and silage.
- Because the stored energy used for regrowth is in the lower stem base located above the soil surface, some animals graze the plants too close for rapid regrowth. Requires good control of defoliation times and height to favor plant survival.
- Orchardgrass usually persists 3–5 years in mid-south, but may last up to 10 years in higher elevations and in the northern transition zone. Makes a good companion plant with slow growing native grasses because it will provide relatively quick cover during the planting year and as the native grasses become more competitive, orchardgrass thins.

Toxicity

- None recorded.

Reference

- Orchardgrass [4,7,12,48,49,53,58,72]

Panicum, Fall—*Panicum dichotomiflorum* Michx.

Fall panicum seeds

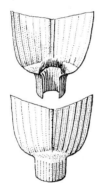

Collar region of fall panicum

Fall panicum plant with seedhead.

Seed

- Size: grain 1.5—1.8 mm long; lemma and palea 2.2—2.8 mm long
- Shape and texture: elliptical, somewhat compressed with one face convex and the other nearly flat; lemma and palea are smooth and shiny with prominent midrib, under high magnification, the surface appears striate
- Color: light to dark olive, the nerves, apex, and base are paler.

Fall panicum plant

Leaf sheath and leaf blade—Vernation: Leaf rolled within the leaf sheath; **Leaf sheath**: round, smooth, split with overlapping hyaline margins; **Leaf blade** is prominent; with large round, smooth sheaths that are often bent at the nodes, white mid-rib on the leaf blade; **Collar**: broad; **Auricle**: absent; **Ligule**: membranous, fringed with hairs.

Stolon/Rhizome/Roots: No stolons; no rhizomes; has fibrous roots. A primary identifying characteristic of this grass weed is that the stem bends in a zigzag manner at each joint, nodes are swollen.

Inflorescence: Flower head is much-branched panicle with purple tinged spikelets.

Panicum, fall—*Panicum dichotomiflorum* Michx.

Other common names: Spreading witchgrass, smooth witchgrass, western witchgrass, sprouting crabgrass
Family: Gramineae/Poaceae
Life cycle: Warm-season annual
Native to: Europe

Distribution and Adaptation
- Primarily found in the warm humid and cool humid regions, but also in the plains states.
- Found on a wide range of soil types that range from well to poorly drained.
- Often found in new plantings of forages, thin stands on damp soils, roadsides, and noncropland sites.

Morphology/Growth Pattern
- Plant grows about 5 ft (1.5 m) in height; leaves smooth, broad; grows from a bent or twisted base, appears "Zigzag" in growth due to the binding at the nodes; open and wide panicle, seedhead develops a purple tint when mature.
- Reproduces by seed.

Use and Potential Problems
- Considered an undesirable species in forage systems.
- Makes high quality and palatable hay when harvested prior to early head stage.
- Often very competitive in new spring plantings, especially of native grasses in the eastern United States.
- Important food source for ducks and other birds.
- Reproduces by seed.

Similar Species
- *Witchgrass* (*Panicum capillare* L.) resembles fall panicum. Fall panicum is easily distinguished from witchgrass by its zigzagging appearance of the stem and by lack of hairs. The stems, leaf sheaths, and leaf blades of witchgrass are densely hairy.

Toxicity
- Saponins and other photosensitization agents.

Reference
- Fall panicum [8,9,12,17,24,44,53,58,66,72,76]

Quackgrass—*Elymus repens* (L.) Gould

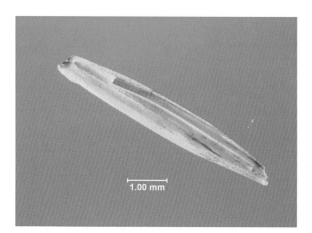

Quackgrass seed

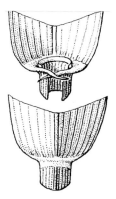

Quackgrass collar region

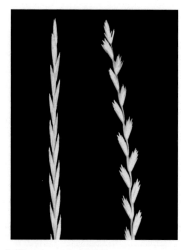

Quackgrass (left); p. ryegrass (right)

Seed
- Size: 11–14 mm long, 2–5.5 mm wide
- Shape and texture: elliptical, broadest below the middle of the seed; "tooth" on edges of palea; both the palea and the lemma are thin and firm
- Color: straw colored, caryopsis is dark

Quackgrass plant and exposed rhizomes

Leaf sheath and leaf blade—Vernation: Leaf rolled in the bud-shoot; **Leaf sheath**: rounded, mostly hairy, split with overlapping hyaline margins; **Leaf blade**: 5–12 mm wide; rough, lower sheaths; hairy, upper sheaths smooth, not compressed, often constricted near the tips; **Collar**: mostly medium broad, may be divided, sometimes minutely hairy; **Auricle**: long and clasps the stem; **Ligule**: membranous, mostly 0.2–0.4 mm long but occasionally 1 mm, truncate to rounded, may be finely toothed or ciliate.

Stolon/Rhizome/Roots: No stolons but strong rhizomes; rhizome tips are yellowish and sharp-pointed; fibrous root systems.

Inflorescence: Spike type seedhead; alternate spikelets with the broad side against the axis of spike unlike ryegrass; white knob at the base of spike; two empty glumes unlike ryegrass with only one empty glume; glumes stiff with five to eight nerves.

Quackgrass—*Elymus repens* (L.) Gould

Other common names: Couch grass, twitchgrass, quick grass, scotch-grass, wheat-grass, devil's grass, knot-grass, and shelly grass
Family: Gramineac/Poaceae
Life cycle: Cool-season perennial
Native to: Europe and Western Asia

Distribution and Adaptation
- Found mostly in the cool humid region but also in the transition zone, especially the northern half of the area.
- Grows in a variety of soils from clay to sandy loams and peat.
- Can withstand high quantities of salt and alkaline conditions as well as flooding.
- Grows in fine-textured soils with a neutral to slightly alkaline soil pH (6.5–8.0) and has been observed in sandy acidic soils.
- Fairly drought tolerant.

Morphology/Growth Pattern
- Forms a very thick sod; spreads by white, long-lived slender rhizome.
- Depending on soil type and environmental conditions, it can grow up to 4 ft (1.2 m) tall.
- Has been reported that one quackgrass plant produced 5 ft (1.5 m) of rhizomes and 206 shoots.
- Reproduces by seed and rhizomes.

Use and Potential Problems
- May be used for pasture, hay, or silage.
- Nutritive value would be similar to other cool season grasses if harvested at similar ages.
- Found in crops, hay and pasture fields, gardens, roadsides, turfgrass, lawns, nurseries, landscapes, and disturbed sites.
- Considered an excellent soil binder.
- Also used for reclaiming nutrients, such as nitrogen, from sewage effluent sprayed on vegetation.
- Considered one of the three most noxious weeds in the United States.
- Quackgrass also serves as a host for various pests.
- The seeds are a serious contaminant in cultivated forage seed, especially in ryegrasses and tall fescue seeds due to similarities in seed size and shape.

Toxicity
- Not a toxin.

Similar Species
- Although the clasping auricles of quackgrass have been used as a key identifying characteristic, it is hard to distinguish this grass from *annual ryegrass* (*Lolium multiflorum* L.), and *perennial ryegrass* (*Lolium perenne* L.), both having auricles. However, none of these grasses have rhizomes like quackgrass and they grow in clumps.

Reference
- Quackgrass [1,7,12,17,24,34,39,44,48,53,56,58,61,72,76]

Redtop—*Agrostis gigantea* Roth (Commonly Referred to as *Agrostis alba*)

Redtop seeds

Seed
- Size: approximately 1.5 mm long, 0.5 mm wide
- Shape and texture: lemma oblong or lance-shaped, round or slightly flattened on the back; palea tapered from near the base of the seed
- Color: light tan
- Seeds per pound: 5,000,000

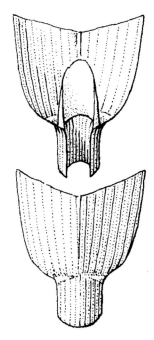

Redtop collar region

Leaf sheath and leaf blade—Vernation: Leaf rolled in bud-shoot; **Leaf sheath**: round, smooth, split with overlapping hyaline margins; **Leaf blade**: up to 10 mm wide, rough above and below, margins rough, narrow leaves and fine stems; mostly basal, hairless and slender; **Collar**: narrow to broad, divided; sometimes minutely hairy; **Auricle**: absent; **Ligule**: membranous, about 5 mm long, rounded to sharp pointed, may be hairy on back.

Stolon/Rhizome/Roots: Rhizomatous; occasionally find stolons.

Inflorescence: Oblong to pyramidal, red to purplish panicles; spikelets, each containing a single floret; glumes exceeding the single floret are often purple to red.

Redtop—*Agrostis gigantea* Roth (commonly referred to as *Agrostis alba*)

Other common names: Couch grass, red top bentgrass, black bent
Family: Gramineae/Poaceae
Life cycle: Cool-season perennial
Native to: Europe

Distribution and Adaptation
- One of the most widely adapted grasses in the United States. More often found in the cool humid and transition zones, however, frequently found in the northern half of the warm humid zone.
- Common around bogs, springs, seeps, ditches, and stream banks.
- One of the best wetland grasses; remains under water for short periods without damage, yet it adapts to dry conditions on acid or alkaline soils.
- Grows in low-fertility, acid, clay, loamy, and sandy soils.
- Good tolerance to drought and poorly drained soils.
- Poor tolerance to traffic, high temperatures, and shade.
- On most soils not as productive as tall fescue or smooth bromegrass.

Morphology/Growth Pattern
- Sod-forming grass because of aggressive rhizome growth.
- May reach 2−3 ft (0.60−0.90 m) height but more often is less than 2 ft (0.60 m) tall, especially prior to head emergence.
- Grows quickly and vigorously therefore provides quick cover for critical areas, erosion control.
- Makes most of its growth in spring and late autumn.
- Reproduces by seed and rhizomes.

Use and Potential Problems
- Used in pasture mixtures, for grazing and hay use.
- Has low palatability, but animals will eat it.
- Also used for quick soil cover for erosion control on critical areas and temporary cover in turf.
- Used to revegetate reclaimed surface mined land.
- Used for nesting cover by upland birds.

Toxicity
- Ergot alkaloids.

Reference
- Redtop [7,12,34,39,49,56−58]

Rescuegrass or Prairie Grass—*Bromus unioloides* (Willd.) H.B.K. (syn. *Bromus catharticus* Vahl & *Bromus willdenowii* Kunth)

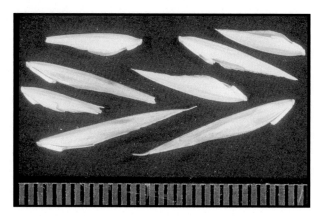

Rescuegrass seeds

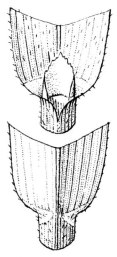

Collar region of rescugrass

Rescuegrass collar region

Seed

- Size: lemma 15−16 mm long, 3 mm wide
- Shape and texture: lemma laterally flattened and sharply keeled, lemma tapers to a point, lemma completely curved around palea
- Color: tan
- Seed per pound: 52,000

Rescuegrass plant (Chris Teutsch)

Leaf sheath and leaf blade—Vernation: Leaf rolled in bud-shoot; **Leaf sheath**: strongly compressed, can be nearly glabrous to densely pubescent with fine, straight, spreading hairs, closed, forming a tube around the stem; **Leaf blade**: flat, blades 20−30 cm long and 9.5−13 mm wide, rough or sparsely hairy; **Collar**: medium broad to broad, divided; **Auricle**: absent; **Ligule**: membranous, 3−6 mm long, rounded, sometimes toothed, often torn, hairy on the back

Stolon/Rhizome/Roots: No stolon or rhizome, fibrous root system.

Inflorescence: Branching panicle, nodding with strongly flattened (condensed) spikelets; spikelets flat and green which turns straw yellow when ripe and dry; the spikelets overlap forming V's with short or no awn; lemmas mostly awnless.

Rescuegrass or Prairie grass—*Bromus unioloides* (Willd.) H.B.K. (syn. *Bromus catharticus* Vahl & *Bromus willdenowii* Kunth)

Other common names: Prairie bromegrass, matua is a variety, but commonly used as name
Family: Gramineae/Poaceae
Life cycle: Cool-season, short-lived perennial; but often performs as an annual
Native to: Prairie grass (Pampas of South America); rescuegrass (Argentina and Uruguay)

Distribution and Adaptation
- Primarily found in the transition zone, the northern areas of the warm humid zone, and the southern area of the cool humid zone.
- More drought resistant and continues to grow later in the fall than most cool-season forage grasses.
- Adapted to well-drained, high fertile soils. Grows best in soils with a pH of 6.0−7.0.
- Grows well in sandy, drier soils that often limit other cool season grass species.
- Tolerate heat, and droughty soils; intolerant of poorly drained soils.
- Grows actively in late fall/early winter; are susceptible to winter kill in areas with severe winters and no snow cover.
- Often found as a volunteer plant in pastures, roadsides, and other areas with productive soils.

Morphology/Growth Pattern
- An erect growing plant, typically 2−3 ft (0.60−0.90 m) tall, including the inflorescence; have bunch-type growth habit.
- Produce seedheads following each grazing or harvesting.
- Produces stems with elevated growing points or joints without flower production. The elongation of the stem after defoliation elevates the growing point above soil level even though the terminal bud does not become reproductive. Frequent defoliation of these species can eventually reduce stands and yields because of the fact that terminal buds (meristems) are removed while leaf area for photosynthesis is reduced and carbohydrates in the stem base are not sufficient to support rapid regrowth.
- The production of elevated growing points makes these species less suited for grazing compared with orchardgrass and tall fescue, which remain vegetative during summer and fall with the growing point buds protected near soil level.
- Most productive early spring, late winter and fall, and has the capacity for greater summer production when irrigated.
- Large amounts of seed produced throughout the year ensure annual reseeding.

Use and Potential Problems
- Suited for grazing, hay, or silage.
- Highly palatable at all stages of growth, seem to be more palatable at advanced stages than other commonly grown forage grasses.
- Because it is very palatable, it does not persist in mixtures with less palatable species, because animals tend to bite it frequently and close to the soil leaving few leaves and little reserve energy storage.
- The most common disease is powdery mildew (*Blumeria* spp. (*Oidium*) of prairie grass is head smut (*Ustilago bullata* Berk)
- May become a weed in lawns and disturbed sites and also used for grazing and hay.

Toxicity
- Powdery mildew.

Reference
- Rescuegrass [2,7,12,32,56,57,73]

Rye, Grain—*Secale cereale* L.

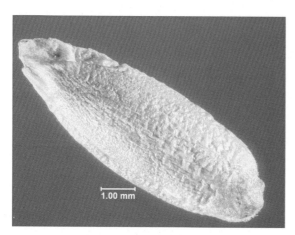

Rye seed

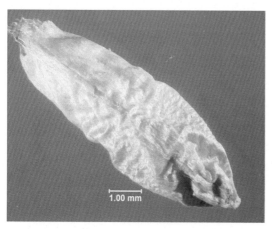

Triticale seed

Seed
- Shape and texture: long-pointed germ end; seed coat has small "scales"; triticale seed is much more wrinkled
- Color: greenish-gray seed color
- Seeds per pound: 18,000 (triticale 16,000)

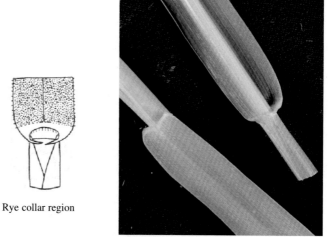

Rye collar region

Rye collar region

Cereal rye head

Leaf sheath and leaf blade—Vernation: Leaf rolled in the bud-shoot; **Leaf sheath**: round, smooth, split with overlapping hyaline margins; **Leaf blade**: 8–18 mm wide, rough above, smooth and keeled at least near base, margins rough and hyaline; **Collar**: mostly broad; **Auricle**: small; **Ligule**: membranous, 1–1.6 mm long, rounded, jagged, may be ciliate.

Stolon/Rhizome/Roots: No stolons or rhizomes, fibrous root systems.

Inflorescence: Long and narrow, crowded spike with thin stem; partly exposed kernels; teeth and tiny hairs on glumes; two glumes per spikelet; pubescent near the spike; 40–50 spikelets on the rachis.

Rye, grain—*Secale cereale* L.

Other common names: Cereal rye, winter rye, grain rye, cover-crop rye
Family: Gramineae/Poaceae
Life cycle: Cool season, annual
Native to: Southwestern Asia/Northern Europe

Distribution and Adaptation
- Grown in every state in the United States, but as a forage it is grown widely in the warm humid and transition zones. The leading states in rye seed production include South Dakota, Georgia, Nebraska, North Dakota, and Minnesota.
- Can be grown in a wider range of environmental conditions than any other small grain.
- Rye can grow on more infertile, sandy, or acid soils than other cereal grains.
- Best adapted to fertile, well-drained sandy loams having a pH of 5.6 or higher rather than on heavy clay soils.
- Fairly tolerant to drought conditions.

Morphology/Growth Pattern
- An erect annual grass with flat leaf blades.
- Grows from 2 to 4 ft (0.60−1.2 m) tall.
- Of the most cold tolerant plants and usually makes more growth at cooler temperatures than other small grains, with the possible exception of triticale. However, does not survive the winter if standing on wet areas where ice sheets are formed.
- In the warm humid and transition zones it forms seedheads from 2 to 4 weeks earlier than other small grains.
- Reproduces by seed.

Use and Potential Problems
- Used for grazing, haylage, or green chop prior to heading.
- Used extensively as a winter cover crop to control erosion, weed suppression, pest suppression, and as a green manure crop.
- Used for biomass/organic matter source and nutrient scavenging during cool seasons.
- Also used for alcoholic beverages, food, and seed.
- In the United States, rye is usually mixed with 25%−50% wheat flour for bread making.
- Straw used as packing material for nursery stock, bricks and tiles, for bedding, paper manufacturing, archery targets, and mushroom compost.
- Use for weed control due to its allelopathic effect on other plants especially in no-till culture.

Similar Species
- Similar plant triticale is a cross between wheat and rye, thus morphological characteristic of this plant combines the morphological characteristics of both wheat and rye.

Toxicity
- Photosensitization agents.

Reference
- Rye [4,7,12,33]

Ryegrass, Annual—*Lolium multiflorum* Lam.

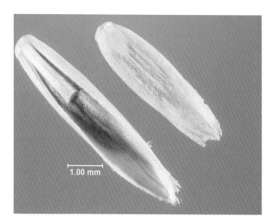

Annual ryegrass seeds

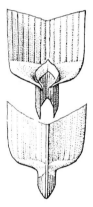

Collar region of annual ryegrass

Annual ryegrass seedhead *Source: Sam Doak*

Seed

- Size: 5−8 mm long, 1−1.5 mm wide.
- Shape and texture: seed has a tapering rachilla wide at base and cut straight across at top; seed is like tall fescue except rachilla is flat, rectangular, and not knobbed like tall fescue
- Color: tan
- Seeds per pound: 250,000

Annual ryegrass plant *Source: Sam Doak*

Leaf sheath and leaf blade—Vernation: Leaf rolled in the bud-shoot; **Leaf sheath**: distinctly flattened to almost round, smooth, closed to near top; **Leaf blade**: shiny, glabrous, long, glossy; **Collar**: collar region is narrow, hairless, and yellowish- to whitish-green; **Auricle**: clasping, prominent, tapering to a narrow point; **Ligule**: membranous, up to 1−2 mm long; rounded or toothed at the tip.

Stolon/Rhizome/Roots: Erect culms often tinged red at the base; no stolons or rhizomes; plant has extensive fibrous root system.

Inflorescence: The inflorescence is composed of about 35−40 sessile spikelets arranged alternately, with 10−20 fertile florets per spikelet. Awns are usually present and attached to the lemma.

Ryegrass, annual—*Lolium multiflorum* Lam.

Other common names: Italian ryegrass, Australian rye, English ryegrass, winter rye, rye grass
Family: Gramineae/Poaceae
Life cycle: Cool-season annual
Native to: Southern Europe, North Africa, and southwest Asia

Distribution and Adaptation
- About 90% of annual ryegrass (one million hectares) is grown in the southeastern United States (warm humid zone) for winter pasture.
- Adapted to a wide range of climate and soil conditions, best grown under cool and moist climates.
- Tolerates cool soil temperature better than most other winter annual crops; grows well on sandy soils but performs best on heavier soils such as clay or silty soils with adequate drainage.
- Tolerates extended wet soil conditions as well as short flooding periods.
- Tolerates pH from 5 to 8 with optimum between 6 and 7.
- Moderately shade tolerant.

Morphology/Growth Pattern
- Bunch type of grass, with erect culms, grows to a height of 2–5 ft (0.60–1.5 m).
- Seedling growth is very vigorous and competitive with associated plants.
- Produces most of its growth in late autumn and early spring.

Use and Potential Problems
- Used extensively for grazing and to a lesser extent for hay or silage.
- Palatability and nutritive value are best of the forages used by animals.
- Its vigorous growth habit makes it a serious pest during establishment of other slower growing species.
- Considered a weed in small grains, turf, nurseries, and other perennial crops.
- Often used as a nurse crop to provide quick soil cover to prevent erosion; may dominate mixture and result in less than desirable stands of the permanent species.
- Used as a cover crop, and in critical area plantings such as waterways or riparian areas subject to flooding because it tolerates wet soils and temporary flooding.
- Reproduces by seed.

Similar Species
- Annual ryegrass seedheads are similar to *perennial ryegrass* (*Lolium perenne* L.) except that *annual ryegrass* spikelets have awns, additionally perennial ryegrass has more leaves in lower parts of the plant canopy, its collar and blade are more narrow, and lemmas are awnless.

Toxicity
- Nitrate poisoning.

Reference
- Annual ryegrass [1,4,7,12,29,30,32,33,44,48,49,72,76]

Ryegrass, Perennial—*Lolium perenne* L.

Perennial ryegrass seeds

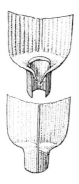

Collar region of perennial ryegrass

Perennial ryegrass seedhead (NCSU)

Seed
- Size: 5.0−8.0 mm, and width at the midpoint is 1.0−1.5 mm long
- Shape and texture: seed has a tapering rachilla wide at base and cut straight across at top; seed is like tall fescue except rachilla is flat, rectangular, and not knobbed-like tall fescue
- Color: tan
- Seeds per pound: 250,000

Perennial ryegrass plant *Source: Sam Doak*

Leaf sheath and leaf blade—Vernation: leaves folded in the bud-shoot; **Leaf sheath**: distinctly flattened to almost round, smooth, closed to near top; **Leaf blade**: up to 6 mm wide, smooth above and below, prominently ridged on the upper surface, glossy and keeled below, margins slightly rough; **Collar**: collar region is narrow, hairless, and yellowish to whitish-green; **Auricle**: small, soft, and claw-like, often tapering to a narrow point; **Ligule**: thin-membranous, 0.5−1.5 mm long; rounded or toothed at the tip.

Stolon/Rhizome/Roots: Bunch type, no stolon or rhizome, highly branched shallow root system; produces adventitious roots from the basal nodes of tillers.

Inflorescence: Has long narrow spikelets attached directly to rachis and is placed edgewise along the main stem; spikelets flattened *without awns (annual ryegrass have awns)*; only one glume per spikelet (quackgrass has two glumes per spikelet); several florets per spikelet (6−10 seeds). Often confused with quackgrass. However, quackgrass is a rhizomatous, and the spikelets of quackgrass are arranged with the broad side against the axis of spike. Quackgrass also has two empty glumes unlike ryegrass with only one empty glume.

Ryegrass, perennial—*Lolium perenne* L.

Other common names: Ryegrass, English ryegrass, crested ryegrass
Family: Gramineae/Poaceae
Life cycle: Cool-season perennial
Native to: Southern Europe, North Africa, and southwest Asia

Distribution and Adaptation

- Mostly grown as a forage in the northeast area of the cool humid zone. It will perform well for one to two seasons in the northern sections of the transition zone, but annual ryegrass is probably a better choice for those areas. It is not a good alternative in the warm humid zone.
- Grows best on fertile, well-drained soils; will grow on soils with pH of 5–8, but production is best at pH 6–7.
- Best adapted to cool, moist climates.
- Less winter hardy, and less persistent than most cool-season grasses (Tall fescue (*Festuca arundinacea*), orchardgrass (*Dactylis glomerata* L.), timothy (*Phleum pratense* L.), smooth bromegrass (*Bromus inermis* L.), and Kentucky bluegrass (*Poa pratensis* L.).
- Perennial ryegrass is less drought tolerant than orchardgrass (*D. glomerata* L.).

Morphology/Growth Pattern

- Bunch type of grass, with erect culms, grows to a height of 2–5 ft (0.60–1.50 m).
- Flowering is induced following a period of cool temperature and short days, therefore regrowth following harvests in the heading stage is mostly leaf.
- Often behaves as an annual the southern area of the transition zone and the warm humid zone.
- Reproduces by seed.

Use and Potential Problems

- Used for hay, silage, grazing, and turf.
- One of the best quality forage grasses in the world.
- Due to its extensive root system and high response to fertility, it is used for nutrient recycling systems where stands persist for several years.
- Useful as wildlife feed; for geese, coots, widgeons and other ducks, wild turkeys, rabbits, deer, and elk.
- In the United States expanded use of perennial ryegrass is limited by lack of adaptability to high temperature, high humidity, and disease pressure.
- Fungal endophytes often is found in varieties to increase pest resistance and persistence for turf purposes, but these endophytes may adversely affect animal health (ryegrass staggers) in grazing animals.
- In many areas of the upper and lower south, stands of perennial ryegrasses are eliminated in a couple of years by *Rhizoctonia solani* Kuehn.

Toxicity

- Nitrate poisoning.

Reference

- Perennial ryegrass [1,4,7,12,29,30,32,33,44,48,49,72,76]

Side Oats Grama—*Bouteloua curtipendula* (Michx.) Torr.

Side oats grama seed

Seed
- Size: lemma and palea 3.5–6.5 mm long
- Shape and texture: lemma three-nerved, glabrous or pubescent, acute or short-awned; paleas whitish, usually glabrous
- Color: tan
- Seeds per pound: 112,000

Side oats grama plant

Leaf sheath and leaf blade—Vernation: Leaf rolled in the bud-shoot; **Leaf sheath**: usually shorter than the internodes, fine parallel lines or ridges on sheath, glabrous to somewhat hairy; **Leaf blade**: numerous, flat, with stiff hairs along the margins, especially near the base, leaves mostly basal; as the grass matures, the leaves become curled and whitish-brown; **Collar**: the collar is often hairy on the margins; **Auricle**: no auricle; **Ligule**: very short fringed membrane, truncate.

Stolon/Rhizome/Roots: No stolons, produces short, scaly rhizomes; fibrous root system.

Inflorescence: Panicle 7.5–40.5 cm long, composed of many one-sided spikelets, projecting downward in two rows along one side of the culm; one-sided spikelets each contain a few to several spikelets. The one-sided spikelet arrangement gives it the appearance of "the flags on a ship's mast."

Side oats grama—*Bouteloua curtipendula* (Michx.) Torr.

Other common names: None known
Family: Gramineae/Poaceae
Life cycle: Warm-season perennial
Native to: United States

Distribution and Adaptation
- Found but not widely evident in the transition and cool humid zones.
- Occurs naturally in limestone glades, shallow soils, upland prairies, and savannas.
- Prefers fine-textured soils that are calcareous or somewhat alkaline. Excellent for dry soils, as well as well-drained loams and clays.
- Often found in association with little and big bluestem (*Andropogon* spp.) plant communities.
- Because it is so drought tolerant it often becomes abundant following prolonged dry periods.

Morphology/Growth Pattern
- Plants maintain leaves near the soil surface which provides tolerance to close grazing.
- "Side oats" refers to the small oat-like seeds that hang down uniformly on one side of the stem.
- Grows to 1−3 ft (0.30−0.90 m) tall.
- The short scaly rhizomes often give plants a bunchy appearance.
- Stands are readily established by seeding.
- Has one of the most attractive, showy bright purple and orange flowers of any grass. The seedhead is equally appealing, with small oat-like seeds suspended on one side of the stalk.
- Reproduces by seed and rhizomes.

Use and Potential Problems
- Used for grazing or hay.
- Forage quality and palatability are excellent.
- Tends to provide good soil cover in times of drought and under severe grazing.
- Provides a good food source for foraging wildlife.

Toxicity
- None recorded.

Reference
- Side oats grama [2,7,32,34,39,49,56,68]

Signalgrass, Broadleaf—*Urochloa platyphylla* (Munro ex C. Wright) R.D. Webster (Also Known as *Brachiaria platyphylla* (Munro ex C. Wright) Nash)

Broadleaf signalgrass seeds

Signalgrass collar region

Broadleaf signalgrass leaf, stem and seedhead

Seed
- Size: grain 2 mm long; lemma and palea 2.5−3 mm long
- Shape and texture: broadly elliptical
- Color: yellowish in color

Broadleaf signalgrass plant

Leaf sheath and leaf blade—Vernation: Leaf rolled in the bud-shoot; **Leaf sheath**: often maroon-tinged and hairy throughout; **Leaf blade**: overall short and wide, 4−15 cm long, 6−15 mm wide, widest near the base and tapering to the apex; no hairs except on margins and in the collar region; often partly folded or creased near the tip; **Collar**: hairy; **Auricle**: absent; **Ligule**: narrow membrane fringed with hairs, 0.5−1 mm long.

Stolon/Rhizome/Roots: No stolons or rhizomes; fibrous; lower stem nodes often root when soil moisture is good.

Inflorescence: Seedhead is a raceme, ascending 4−8 cm long; spikelets on the seedhead are somewhat flattened in appearance.

Signalgrass, broadleaf—*Urochloa platyphylla* (Munro ex C. Wright) R.D. Webster (also known as *Brachiaria platyphylla* (Munro ex C. Wright) Nash)

Other common names: None reported
Family: Gramineae/Poaceae
Life cycle: Warm-season, annual
Native to: Eastern and southern Africa

Distribution and Adaptation
- Most commonly found in the warm humid zone, especially along the coastal plain regions of the zone.
- Frequently found on fine-textured soils but may be found on many courser textured soils.
- Frequently found in association with crabgrass, goosegrass, yellow and green foxtails, millets and sorghum.

Morphology/Growth Pattern
- Semiprostrate growth habit after it reaches 1.2–1.7 ft (0.37–0.52 m) height.
- Highly branched and bent at the nodes; plants often root at the lower stem nodes, and are commonly seen growing along the ground with some tips ascending.
- Prolific seed producer that volunteers in subsequent years.
- Regrowth dependent on reserves stored in stem base, therefore grazing or clipping less than 4 in. (10 cm) can significantly reduce regrowth rates.
- Reproduces by seed.

Use and Potential Problems
- Often volunteers in pastures, especially summer annual forages.
- Palatability and nutritive value are good for cattle, goats, and horses.
- Occurs mostly on disturbed, moist soils and can often be found in fields and ditches.
- Is a vigorous growing seedling and can be very competitive during establishment of warm season perennial plants.

Toxicity
- None recorded.

Reference
- Broadleaf signalgrass [14,24,32,44,54]

Smutgrass—*Sporobolus indicus* (L.) R. Br. var. *capensis* Engl. also called *Sporobolus poiretii* L.

Smutgrass seeds

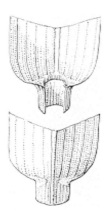

Smutgrass collar region

Smutgrass seedhead

Seed
- Size: grain 0.5−1 mm long, lemma 1.5−2.2 mm long, palea 1.5−1.8 mm long
- Shape and texture: grain flat, oblong; caryopsis rough textured
- Color: grain reddish

Smutgrass collar region

Leaf sheath and leaf blade—Vernation: Leaf rolled in the bud; **Leaf sheath**: smooth, round; **Leaf blade**: flat at the base of plant and becomes rounded, less than 6 mm wide, tapering to a sharp point; **Collar**: divided into two parts by the mid-vein; **Auricle**: absent; **Ligule**: fringe of hairs.

Stolon/Rhizome/Roots: No stolons or rhizomes; fibrous root system.

Inflorescence: Long, narrow, slender, spike-like panicle, often black with smutty fungus; two-flowered, one fertile and sessile; seedhead appears rattail and sometimes is included in the leaf sheath.

Smutgrass—*Sporobolus indicus* (L.) R. Br. var. *capensis* Engl. also called *Sporobolus poiretii* L.

Synonyms: *Sporobolus poiretii* L.
Other common names: Rattail smutgrass
Family: Gramineae/Poaceae
Life cycle: Warm season perennial
Native to: Asia

Distribution and Adaptation
- Found primarily in the warm humid zone, but to a lesser extent in the southern transition zone.
- Commonly found in pastures, turf, meadows and disturbed and over grazed or high traffic areas.
- Often found in pastures that have been severely overgrazed.

Morphology/Growth Pattern
- Smutgrass is a dark green, tufted perennial with erect stems, up to about 3 ft (0.90 m) tall.
- Reproduces by seed.

Use and Potential Problems
- Not readily grazed by livestock or wildlife because it has high tensile strength and is tough to grasp.
- Can become a serious weed problem in many pastures, turf, roadsides, and waste areas.
- Considered to be invasive.
- In the lower south of the warm humid zone up to 30% of pastures are heavily infested with this weed.
- Control of smutgrass includes mowing frequently and close, burning or complete renovation.

Toxicity
- None recorded.

Reference
- Smutgrass [14,32,44,46,48,58,66]

Sorghum/Sudangrass—*Sorghum bicolor* (L.) Moench/*Sorghum bicolor* var. *sudanense*

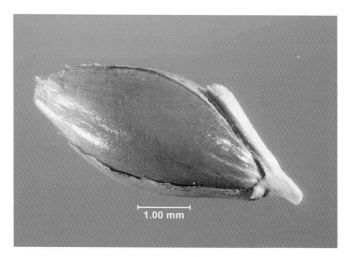

Sweet sudangrass seed (*Sorghum bicolor* var. *sudanense*)

Grain sorghum seeds *Source: SDSU*

Seed (Sweet Sudangrass)
- Size: length 4–4.5, width 2–2.5 thickness 1.5–2 mm
- Shape and color: flattened seed, glossy, no knobs on rachilla as Johnsongrass
- Color: red to purple
- Seeds per pound: Sudangrass 55,000; sorghum-sudangrass hybrids 20,000

Seed (Sorghum)
- Shape and color: round seed with dark undercoat, minor cracked paint; bluish white (hegari); somewhat long, pointed
- Color: seed coat is dull orange with a few dark glumes (orange sorgo)
- Seeds per pound: 13,000–24,000

Sorghum-sudangrass plant

Leaf sheath and leaf blade—Vernation: Leaf rolled in the bud; sorghum-sudangrass leaves are shaped like corn, but are shorter and not as wide us corn, leaf blades are waxy and globerous; **Leaf sheath**: open, smooth; **Leaf blade**: boat shaped, smooth; **Collar**: medium broad to broad; **Auricle**: absent; **Ligule**: long, with fringe of hair.

Stolon/Rhizome/Roots: No stolons or rhizomes, fibrous root system.

Inflorescence: Long open panicle with red glumed seeds, reddish orange at the base blending to straw color at the tip of each glume.

Sorghum/Sudangrass—*Sorghum bicolor* (L.) Moench/*Sorghum bicolor* var. *sudanense*

Other common names: Sorghum, broomcorn, milo
Family: Gramineae/Poaceae
Life cycle: Warm-season, annual
Native to: Africa

Distribution and Adaptation
- Primarily adapted to the Southeastern states.
- Sudangrass and sorghum-sudangrass hybrids are best adapted to well-drained, fertile soils, however, can grow on somewhat poorly drained soils when surface water is drained, drought tolerant but does not tolerate low pH and requires liming if grown on acid soils (pH 6.0—6.5).

Morphology/Growth Pattern
- Forage *sorghum-sudan* is very similar to *corn* (*Zea mays* L.) in the vegetative stage; leaves tend to be more narrow than corn; heavily covered with a white waxy coating that can be rubbed off the leaf sheath; flowers both male and female parts produced in a panicle type head on top of the plant; forage sorghum-sudan can grow 6—10 ft (1.8—3 m) tall. *Sudangrass (Sorghum sudanense)*; smooth, large leaves, erect stem reach a height of 5—7 ft (1.5—2 m); open panicle. Sorghum-sudangrass is tall growing, resembles sudangrass, but has coarser stems, taller growth habit, and higher yielding than either sorghum or sudangrass.
- Reproduces by seed.

Use and Potential Problems
- Sudangrass and sorghum sudangrass can be used for hay or silage. However, hay curing is difficult due to thick stem, stem can be crushed mechanically to reduce drying time.

Similar Species
- *Sorghum-sudangrass* is tall growing, resembles sudangrass, but has coarser stems, taller growth habit, and higher yielding than either sorghum or sudangrass.

Toxicity
- Like sorghum, sudangrass and sorghum-sudangrass hybrids contain *prussic acid* (hydrocyanic acid HCN); sorghum-sudangrass contain higher concentrations of prussic acid than sudangrass; prussic acid can cause poisoning in cattle when young, drought stressed, or frosted forage is grazed. To reduce the incidence of prussic acid poisoning, do not graze young, drought stress, frosted, or damaged plants. Varieties differ considerably in amount of HCN in plant. HCN is destroyed when plant is ensiled or cured as hay. *Nitrate poisoning*: caused by excessive nitrogen fertilization; do not graze slow growing or drought stressed stands.

Reference
- Sorghum-sudan/sudangrass [2,4,7,8,12,15,29—32]

St. Augustinegrass—*Stenotaphrum secundatum* (Walt.) Kuntze

St. Augustinegrass seeds

St. Augustinegrass collar region

St. Augustinegrass *Source: NCSU turf*

Seed
- Size: spikelet 4–5 mm long
- Shape and texture: lemma glabrous and long-pointed, exceed the fertile lemma
- Color: seed coat tan, caryopsis dark brown

St. Augustinegrass *Source: NCSU turf*

Leaf sheath and leaf blade—Vernation: Leaf folded in the bud; **Leaf sheath**: strongly compressed and keeled, slightly hairy along edges and toward top; **Leaf blade**: flat, 4–10 mm wide, smooth on both surfaces except near the ligule, with blunt tip; **Collar**: broad, continuous, narrowed to form a short stalk at the base of the blade; **Auricles**: absent; **Ligule**: a fringe of hairs, 0.3 mm long.

Stolon/Rhizome/Roots: Stolons and no rhizomes.

Inflorescence: short, one-sided, spike-like raceme 4–15 cm long, terminating the stem and each flowering branch.

St. Augustinegrass—*Stenotaphrum secundatum* (Walt.) Kuntze

Other common names: Buffalo grass, St. Augustine grass
Family: Gramineae/Poaceae
Life cycle: Warm-season, perennial
Native to: Gulf of Mexico region, the West Indies, and Western Africa

Distribution and Adaptation
- Native to Australia, West Indies, and southern Mexico.
- Adapted to a wide range of environmental conditions, grows best in moist, well-drained, sandy, slightly acidic soils of moderate to high fertility.
- Poor cold tolerance therefore best adapted to coastal areas with mild winters; tolerant of short dry period; withstands flooding; thrives in shaded areas.

Morphology/Growth Pattern
- Extensively branching perennial, numerous noded-stolon with long internode with short leafy branches; no rhizomes.
- The stem reaches 2.4−15.7 in. (6−40 cm) or more, much branching from numerous nodes.
- A course textured, aggressive, stoloniferous, warm-season perennial grass.
- Propagated by stolons, do not produce seeds.

Use and Potential Problems
- Widely used as a lawn grass where adapted.
- Good for erosion control.
- Loses palatability rapidly thus not good for use in livestock feed.
- Problems associated with this grass include coarseness, low productivity and its lack of seed production.

Toxicity
- None recorded.

Reference
- St. Augustinegrass [34,38,59,67,70,71]

Switchgrass—*Panicum virgatum* L.

Switchgrass seeds

Seed

- Size: 1.2 mm long; lemma and palea 3.2 mm long, 1.5 mm wide
- Shape and texture: teardrop shaped, shiny; palea and lemma sometimes mottled with dark gray
- Color: light to dark gray
- Seeds per pound: 390,000−400,000

Leaf sheath and leaf blade—Vernation: Leaf rolled in

Switchgrass in clump also showing stem base and root system

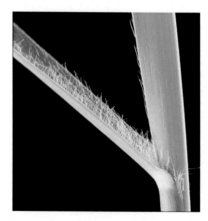

Switchgrass collar region

Switchgrass collar region

the bud-shoot; **Leaf sheath**: round, smooth, split, as long as or longer than the internodes, often reddish or purplish at the base; **Leaf blade**: flat, 6−12 mm wide, mostly hairy above at least near base and distinctly veined; **nest of hair on blade** where the blade attaches to the sheath; **Collar**: broad, may be hairy on margins; **Auricle**: absent; **Ligule**: a fringe of hairs 2−3 mm long, fused at base; a dense ring of hair.

Stolon/Rhizome/Roots: No stolons; short rhizomes; fibrous roots; deep rooted, roots can grow up to 3 m in depth.

Inflorescence: Long open panicle, ovate to pyramidal in shape; spikelets on the end of the long branches; one fertile floret/spikelet; glumes over 12 mm the length of the spikelet; appears to have three glumes, often purple during the flowering and early seed formation stages.

Switchgrass seedhead

Switchgrass—*Panicum virgatum* L.

Other common names: Switchgrass, prairie switchgrass, tall panic grass, water panicum
Family: Gramineae/Poaceae
Life cycle: Warm-season perennial
Native to: United States

Distribution and Adaptation
- Found growing in most states in eastern United States.
- Best adapted to moist, fertile soils, tolerates droughty, infertile, eroded soils.
- Tolerant of moderate soil salinity and acidity, soil pH from 4.5 to 7.6.
- Tolerant of flooding up to several days.
- Tolerant of droughts and partial shade.

Morphology/Growth Pattern
- Grows in large clumps, erect, coarse, and 3 ft (0.90−2.2 m) in height.
- Switchgrasses are characterized as two ecotypes: lowland and upland.
  - Lowland ecotypes: coarse-stemmed, glabrous, more rapid and taller growth than upland ecotypes, typically found on moist, fine-textured soils.
  - Upland ecotypes: fine-stemmed, varying leaf blade pubescence, semidecumbent, commonly found in courser textured and well-drained soils.
- Produces tremendous amount of seeds but a large percentage of the seed is not viable, unless aged or wet chilled.
- Color of foliage and seedhead depends on cultivar. A few examples: common types turn pale yellow in the fall; "Shenandoah" has burgundy foliage; "Dallas Blues" has bluish foliage, "Rehbraun" is 3−4 in. (7.6−10.2 cm) tall with red foliage in the late summer.
- Switchgrass is sensitive to low grazing/cutting height due to its growing point being elevated early in vegetative growth and having a high ratio of reproductive to vegetative tillers. Frequent cutting below 8 in. (20 cm) can weaken a switchgrass stand.
- Reproduces by seed and short rhizomes.

Use and Potential Problems
- Good forage production and quality in spring and early summer.
- Excellent erosion control in filter strips, grass hedges, and as ground cover.
- Used as a renewable biomass energy source.
- Provides food (forage and seeds) and season long cover for a variety of wildlife.
- Problems with stand establishment due to dormant seeds (can be reduced by aging or prechilling moist seeds).
- It can be used to add fall color to naturalistic borders and woodland-edge gardens.

Toxicity
- Saponins and other photosensitization agents.

Reference
- Switchgrass [2,4,7,12,29−32,34,40,49,56,57,58]

Teff—*Eragrostis tef* (Zuccagni) Trotter

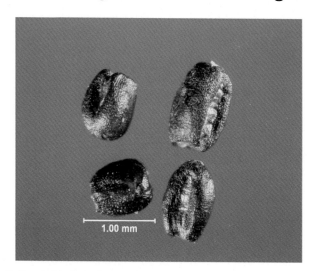

Teff seeds (similar to lovegrass)

Seed

- Size: 0.9−1.7 mm in length, 0.7−1.0 mm in diameter
- Shape and texture: small, round, smooth
- Color: white to dark brown
- Seeds per pound: 750,000−1.4 million

Teff field near Blacksburg, Virginia

Teff—bunch type of growth habit

Leaf sheath and leaf blade—Venation: rolled in the leaf bud; **Leaf sheath**: smooth, glabrous, open, and distinctly shorter than the internodes; **Leaf blade**: slender, narrow, linear; **Auricle**: short auricle, a fringe of hair; **Ligule**: very short and ciliated.

Stolon/Rhizome/Roots: No stolons; no rhizomes; erect stems; shallow, massive, fibrous root system.

Inflorescence: Panicle type of inflorescence ranging from loose to compact; spikelets with 2−12 florets; florets each have lemma, palea, three stamens, an ovary and mostly two feathery stigmas.

Teff—*Eragrostis tef* (Zuccagni) Trotter teff

Other common names: Tef, Williams Lovegrass
Family: Gramineae/Poaceae
Life cycle: Warm-season annual
Native to: Ethiopia

Distribution and Adaptation
- Primarily grown throughout Ethiopia for grain; also grown in South Africa, Australia, and United States for both forage and grain.
- Adapted to environments ranging from drought stress to waterlogged soil conditions.
- Optimum teff production occurs at altitudes of 5906−6890 ft (1800−2100 m); growing season rainfall of 18−22 in. (450−550 mm), and a temperature range of 10−27°C (50−80°F).
- Teff is day length sensitive and flowers best with 12 hours of daylight; longer days will keep Teff at vegetative stage.

Morphology/Growth Pattern
- Fine-stemmed, bunch-grass with large crowns and many tillers. Depending on variety, it can grow 1−4 ft (0.30−1.22 m) in height.
- Teff can emerge in as little as 3 days (water dependant) but typically emerges 5−6 days after planting.
- Teff is very sensitive to close grazing/cutting height. Cutting below 3−4 in. (7.6−10 cm) can weaken the stand since the growing points come off buds along the tiller stem.
- Reproduces by seed.

Use and Potential Problems
- Teff can be used as a forage or grain crop.
- Due to quick emergence and rapid growth, teff can be used as an emergency crop to fill the gap created by the low productivity of cool-season forages during mid-summer or under drought conditions.
- Effective cover crop due to quick emergence and short growing season; excellent erosion control in filter strips, and grass hedges.
- Grown primarily as a cereal crop in Ethiopia. Two grain varieties: white and red/brown. The white variety is preferred over the red/brown varieties.
- The grain is ground into flour, fermented, and made into *enjera*, a sour-dough type flat bread.

Reference
- Ketema, Seyfu. *Teff. Eragrostis teff (Zucc.) Trotter*. Diss. Institute of Plant Genetics and Crop Plant Research, 1997. Rome, Italy: Gatersleben/International Plant Genetic Resources Institute.

Timothy—*Phleum pratense* L.

Timothy seeds

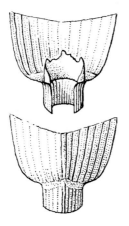

Timothy collar region

Seed
- Size: 1.5 mm long, about the size of bermudagrass
- Shape and texture: oval shaped; may be hulled or unhulled
- Color: naked seed are brown, round usually has thin silvery lemmas attached
- Seeds per pound: 1,330,000

Timothy plant *Source: Wise Co. Virginia*

Leaf sheath and leaf blade—Vernation: Leaf rolled in the bud-shoot; **Leaf sheath**: round, smooth, split with overlapping hyaline margins, fused around the stem; **Leaf blade**: 5−10 mm wide, twisted, ribbed, mostly smooth near base and rough near tip above and below, rough margins; **Collar**: medium broad to broad, may be divided; **Auricle**: absent; **Ligule**: membranous, 2.5−4.7 mm long, rounded, notched near front at least on one side and at the apex.

Stolon/Rhizome/Roots: No stolons; has corms; root system fibrous and shallow, produces a compact swollen bulb at lowest internodes commonly known as haplocorm or corm.

Inflorescence: Dense, narrow, cylindrical, spike-like panicle, rough-textured about 15 cm long.

Timothy—*Phleum pratense* L.

Other common names: Herdgrass, Herd's grass, cat's tail
Family: Gramineae/Poaceae
Life cycle: Cool-season perennial
Native to: Eurasia

Distribution and Adaptation
- Grown in the cool humid and northern areas of the transition zone. Very limited use at elevations above 2000 in the warm humid region.
- One of the most winter-hardy cool-season forage grasses.
- A popular grass where winters are severe and is grown farther north than orchardgrass.
- Grows best on well-drained, deep, fertile, moist clay, or loam soils but can also grow in thin, gravelly, and rocky substrates with adequate moisture.
- Does not persist under drought conditions.

Morphology/Growth Pattern
- Bunchgrass, grows from 1.7 to 3.3 ft (0.50–1 m) height.
- Culms emerge from a swollen or bulblike base (corm) and form large clumps.
- Stems are erect or ascending from a bent base, few vegetative shoots; basal internodes becomes swollen into "bulbs" or "corms."
- Seedhead resembles the foxtails, but unlike timothy, foxtails have hairy ligules and have bristles on spikelets.
- Root system is shallower than orchardgrass and tall fescue.
- Does not require cool temperatures and short day length for floral induction.
- Is a prolific seeder.
- The "corm" stores some carbohydrates that the plant uses for regrowth after harvest or early spring.
- Regrowth rate following defoliation is slower than most other cool season forage species.
- Reproduces by seed.

Use and Potential Problems
- Grazing, hay, silage for all classes of livestock.
- Considered as one of the most important hay grasses in the United States.
- Timothy is widely used for rehabilitation of disturbed sites (mined-land, etc.).
- Provides cover for game birds, small mammals, and waterfowl.
- Has been grown for the extraction of chlorophyll.

Toxicity
- None recorded.

Reference
- Timothy [2,4,7,29–32,34,40,49,53,56–58,72]

Velvetgrass—*Holcus lanatus* L.

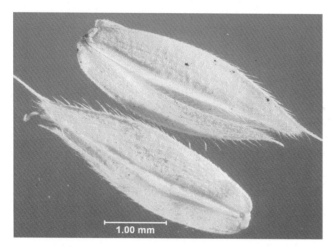

Velvetgrass seeds

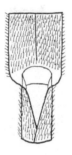

Velvetgrass collar region

Velvetgrass leaf blade and collar region

Velvetgrass seedheads

Seed
- Size: 1.5−2 mm long, 0.5−0.75 mm wide
- Shape and texture: grain elliptical; lemma's ventral side straight, dorsal side arched; smooth, glossy
- Color: grain yellowish; lemma is whitish

Velvetgrassplant *Source: Vivien Allen*

Leaf sheath and leaf blade—Vernation: Leaf rolled in bud-shoot; **Leaf sheath**: distinctly flattened, dense velvety short hair, pink veined, split with overlapping margins; **Leaf blade**: 5.0 mm wide, flat, dense velvety short hair, sharp pointed, margins with short hair; **Collar**: medium broad, hairy, narrow; **Auricle**: absent; **Ligule**: membranous, rounded, medium tall to 3 mm long, jagged at the top and hairy on the back.

Stolon/Rhizome/Roots: No stolons or rhizomes; bunch type of growth; fibrous root system.

Inflorescence: Soft, plume-like panicle, purplish-brown in color, 5−12 cm long.

Velvetgrass—*Holcus lanatus* L.

Other common names: Yorkshire fog, common velvetgrass
Family: Gramineae/Poaceae
Life cycle: Cool-season perennial
Native to: Europe

Distribution and Adaptation
- Commonly found in the transition zone and the southern two-third of the cool humid zone. Occasionally found in the northern one-third of the warm humid zone.
- Found in low producing pastures, old fields, meadows, and roadsides.
- Often grows on damp, rich soils in open areas.
- Tolerates moist, poorly drained and acidic soils. Also tolerates drought.

Morphology/Growth Pattern
- Recognizable from a distance by its soft gray-green color; grows up to 1.5−3 ft (0.46−0.90 m).
- Seeds are produced abundantly and dispersed by wind.
- Reproduces by seed.

Use and Potential Problems
- Considered a weed in low-maintenance pastures, turf, and orchards.
- Low production potential.
- Volunteer in pastures, seldom planted.

Toxicity
- None recorded.

Reference
- Velvetgrass [14,44,48,49,53,56,58,66,76]

Vernalgrass, Sweet—*Anthoxanthum odoratum* L.

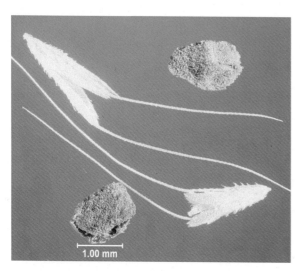

Sweet vernalgrass seeds

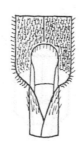

Sweet vernalgrass collar region

Sweet vernalgrass plant

Sweet vernalgrass seedhead

Seed
- Size: 1−1.5 mm long, 0.5−0.7 mm wide, 0.3 mm thick
- Shape and texture: ovate-acuminate, rough
- Color: shiny and golden brown

Leaf sheath and leaf blade—Vernation: Leaf rolled in the bud-shoot; **Leaf sheath**: round, smooth, frequently hairy near top, split with overlapping hyaline margins; **Leaf blade**: flat, edges rough, sharp pointed, mostly short hairs, margins with short hairs; **Collar**: narrow to broad, mostly divided, long hairs at the edges/margins; **Auricle**: absent; **Ligule**: membranous to as much as 9 mm long, mostly rounded, toothed/jagged.

Stolon/Rhizome/Roots: No stolons or rhizomes; tufted stems; a fibrous root system.

Inflorescence: A compressed spike-like panicle with showy yellow flowers; spikelets with awns and a single seed per spike.

Vernalgrass, sweet—*Anthoxanthum odoratum* L.

Other common names: Spring grass, vanilla grass
Family: Gramineae/Poaceae
Life cycle: Cool-season perennial
Native to: Europe

Distribution and Adaptation
- Found primarily in the warm humid and transition zone and the southern areas of the cool humid zone from Illinois, Arkansas, and Louisiana eastward with the exception of Florida.
- Grows primarily in poor soils; thriving best in moist sandy loams.
- Grows on all acidic, neutral and basic soils; does poorly under dry or wet soil.

Morphology/Growth Pattern
- Low growing, tufted perennial with soft green leaves; plant usually under a foot (0.30 m) tall.
- One of the earliest flower grasses in the spring.
- The scented flowers are hermaphrodite (have both male and female organs) and are pollinated by wind.
- Often identified by its sweet aroma; plant becomes more aromatic as it dries or is crushed.
- Reproduces by seed.

Use and Potential Problems
- Found in low performing pastures and hay fields, roadsides, landscaping borders, lawns, meadows, and sand dunes.
- Commonly considered a weed because of low yields and poor nutritive value.
- Produces a lot of pollen and becomes a major allergin.
- Frequent close grazing or mowing will make it less persistent.

Toxicity
- Coumarins, formation of dicoumarol.

Reference
- Sweet vernalgrass [12,14,44,48,49,53,56,58,66,76]

Wheat—*Triticum aestivum* L. subsp. *aestivum*

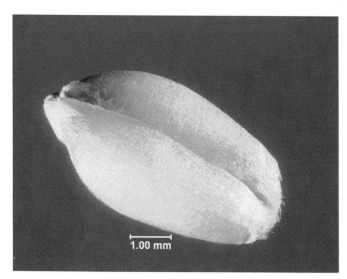

Soft red winter wheat seeds

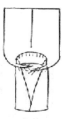

Wheat collar region

Wheat head *Source: Stephen Harrison*

Seed (soft red winter wheat)
- Size: short to medium kernels, 6 mm long
- Shape and texture: soft endosperm; round cheeks, "Open" crease
- Color: orange color; prominent brush
- Seeds per pound: 11,000

Wheat field with Dr. Carl A. Griffey, Small Grains Breeding & Genetics, Virginia Tech

Leaf sheath and leaf blade—Vernation: Rolled in the bud-shoot; **Leaf sheath**: round, smooth, split with overlapping hyaline margins; **Leaf blade**: 7−11 mm wide, smooth near base and rough near tip, margins smooth and hyaline; **Collar**: broad; **Auricle**: clasping auricles, which are short, slender and hairy; **Ligule**: membranous, 1−1.6 mm long, rounded, may be ciliate.

Stolon/Rhizome/Roots: No stolons or rhizomes; fibrous root system.

Inflorescence: A crowded, thick spike.

Wheat—*Triticum aestivum* L. subsp. *aestivum*

Other common names: None found
Family: Gramineae/Poaceae
Life cycle: Cool-season annual
Native to: Iraq, Turkey, and Europe

Distribution and Adaptation
- Soft red winter is widely grown in all zones but to a lesser extent in the extreme northern and southern regions.
- Any moderately well-drained or well-drained soils; wheat is less tolerant of soil acidity than rye; wheat is more tolerant of heavy, wet soils than rye or oats.

Morphology/Growth Pattern
- An erect annual grass with flat leaf blades.
- Grows from 2 to 3.5 ft (0.60−1.0 m) tall.
- For grazing it is generally more productive than oats when temperatures near freezing but not as productive as rye or Triticale.
- Winter wheat requires cold temperatures and short days (vernalization) to induce flowering; spring wheat, however, does not require vernalization to induce flowering.
- Reproduces by seed.

Use and Potential Problems
- Used for grain, pasture, hay, and silage.
- High nutritive value and palatability in the vegetative stages of growth.
- Ensiles well from boot to soft dough stages of growth and compares well with other grass silages.
- May be used as a cover crop but not as often as cereal rye.

Toxicity
- None recorded.

Reference
- Wheat [4,7,29−33,56]

Witchgrass—*Panicum capillare* L.

Witchweed seeds

Seed
- Size: 1.1—1.6 mm long, 0.5—0.7 mm wide
- Shape and texture: floret oval; smooth surface, glossy
- Color: greenish-yellow or brown

Witchgrass plant *Source: Fred Fishel*

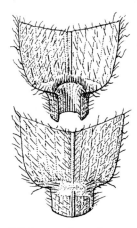

Witchgrass collar region

Witchgrass collar region *Source: Fred Fishel*

Leaf sheath and leaf blade—Vernation: Leaf rolled in the bud-shoot; **Leaf sheath**: somewhat flattened, stiff long hair, split with overlapping margins hairy, usually longer than the inter-nodes; **Leaf blade**: 5—8 mm wide, somewhat wavy, long sparse hair on both sides of the leaf blade, margins smooth and sparse long hair near base; whitish mid rib; **Collar**: continuous, broad, hairy; **Auricle**: absent; **Ligule**: fringe of hairs, about 1 mm long, fused at the base.

Stolon/Rhizome/Roots: No stolons or rhizomes; fibrous root systems.

Inflorescence: A diffuse, dense, panicle that eventually becomes open; two-thirds of the entire height of the plant. The inflorescence is often partially enclosed in the upper sheath and is usually purplish at maturity. Panicle with numerous very fine branches and tiny spikelets at the ends of branches.

Witchgrass seedhead

Witchgrass—*Panicum capillare* L.

Other common names: Old witch-grass, tickle-grass, witches-hair, tumble weed grass, fool-hay
Family: Gramineae/Poaceae
Life cycle: Warm-season annual
Native to: United States

Distribution and Adaptation
- Widespread across the United States, found in all zones described in the book.
- Found in waste and disturbed areas, cultivated fields, and along roads.
- Common in sandy, dry soils as well as moist, fertile soils.

Morphology/Growth Pattern
- Tufted, sprawling or erect, 1–2 ft (0.30–0.60 m) tall.
- Tillers at the base, not at the nodes of elongated shoots.
- Witchgrass resembles fall panicum in the seedling stage. Mature fall panicum blades and sheathes, however, are smooth on both surfaces (no hair).
- Reproduces by seed.

Use and Potential Problems
- Is a common weed of low producing pastures, agronomic crops, gardens, roadsides, and will invade areas where there is little competition.
- Witchgrass provides little forage value, and will rarely be grazed.
- An important food source for quail and other birds.
- Seeds commonly contaminant in forage grass, clover and alfalfa seeds.

Toxicity
- None recorded.

Reference
- Witchgrass [24,34,36,48,49,56,61,66,76]

Zoysiagrass—*Zoysia japonica* Steud.

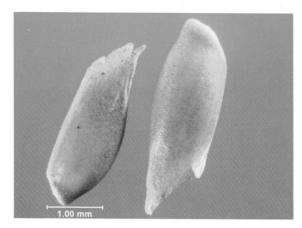

Zoysiagrass seeds

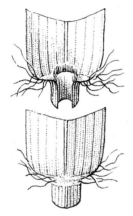

Zoysiagrass collar region

Zoysiagrass hairy stem and collar region *Source: NCSU turf*

Seed
- Size: length 3.5—4 mm, width 1.5 mm
- Shape and texture: broadly ovate, short-pointed or short-awned
- Color: dull, brownish, commonly with a purplish ting

Zoysiagrass plant *Source: Sam Doak*

Leaf sheath and leaf blade—Vernation: Leaf rolled in the bud; **Leaf sheath**: round to slightly flattened, split, glabrous, but with a tuft of hair at the throat; **Leaf blade**: stiff leaf blades with occasional hairs near the base, margins are smooth and blades are sharply pointed; **Collar**: continuous, broad, edges hairy; **Auricle**: absent; **Ligule**: with fringe of hairs like bermudagrass.

Stolon/Rhizome/Roots: Has both stolons and rhizomes, internode of stolons uniform in length unlike bermudagrass which may have variable internode lengths, nodes or stolons generally covered by a tan, husk-like structure; has deep rooted system

Inflorescence: Is a short, terminal spike-like raceme with spikelets on short appressed pedicels.

Zoysiagrass—*Zoysia japonica* Steud.

Other common names: Japanese lawn grass, Korean lawn grass
Life cycle: Warm-season, perennial
Native to: China, Japan, and southeast Asia

Distribution and Adaptation
- In the United States, zoysiagrasses are adapted along the Atlantic coast from Florida to Connecticut and along the Gulf Coast to Texas; also adapted to the transition zone of the United States and in California.
- Can be grown on a wide range of soils from sandy soils to clay.
- Does not perform as well under shade as other shade tolerant grass species; is extremely drought tolerant; but does not tolerate poorly drained soil.

Morphology/Growth Pattern
- The leaf texture of the grass varies from extremely coarse texture to fine-textured types.
- Leaf blades are stiff due to a high silica content.
- A sod forming grass that has both stolons and rhizomes.
- Established by plugging, sodding, or strip sodding.

Use and Potential Problems
- Several species and varieties are used for residential and commercial landscapes, athletic fields, and golf course tees, fairways, and roughs.
- The grass turns brown after the first hard frost and is among the first warm-season grasses to green up in the spring.
- Can be used on golf courses, parks and athletic fields.
- If managed well, zoysiagrass provides a high quality turf, but generally requires a high level of maintenance.
- Is among the most wear tolerant turfgrasses, but its slow rate of regrowth makes the grass slow to recover and thus the grass is not recommended for use in football or soccer fields.
- The major insect problem associated with zoysiagrass is white grubs.
- Troubled by several insects, diseases, and nematodes.

Toxicity
- None recorded.

Reference
- Zoysiagrass [13,49,70,71]

Nonleguminous Forbs

Common Grasses, Legumes and Forbs of the Eastern United States. DOI: https://doi.org/10.1016/B978-0-12-813951-6.00003-0

Basil, Wild—*Clinopodium vulgare* L./*Satureja vulgaris* L.

Wild basil seeds

Wild basil leaves and flowers *Source: Lachlan Cranswick*

Wild basil flowers *Source: Lachlan Cranswick*

Seed
- Size: mericarps 1–1.3 mm long
- Shape and texture: mericarps globose to broadly oblong, smooth
- Color: mericarps dark brown

Wild basil plant

Leaf shape and arrangement—Leaves: Opposite, arranged in pairs, egg-shaped, hairy, rather oval, with short leaf-stalks and slightly toothed margins.

Stolon/Rhizome/Roots: No stolons, rhizomes; stems hairy, square.

Inflorescence: Flowers are dense, terminal head-like cluster or whorled clusters in the axils of the upper leaves; bristly bracts at the base of the flower clusters look "wooly" because of hairs on the calyx and bracts; whorls of pinkish purple flowers.

Basil, wild—*Clinopodium vulgare* L./*Satureja vulgaris* L.

Family: Lamiaceae
Other common names: None reported
Life cycle: Cool-season perennial
Native to: Europe

Distribution and Adaptation
- Found throughout eastern Canada and the adjacent northern United States.
- Mostly found in moist open woods, thickets, roadsides and some in dry grasslands, mostly on calcareous soils.

Morphology/Growth Pattern
- This perennial plant has a square stem and rhizomes.
- Pubescent stem grows up to 9−18 in. (23−46 cm).
- Hairy, aromatic, faint thyme-like odor.
- Reproduces by seed.

Use and Potential Problems
- No known agronomic use.
- This weed species found in old fields, pastures, along roadsides, and at forest borders.

Toxicity
- None recorded.

Similar Species
- The arrangement of the white flowers in cluster, hairy bracts, and the nearly smooth (no tooth) leaf margins are typical of the herb *sweet basil* (*Ocimum basilicum* L.).

Reference
- Wild basil [46−48,61]

Bedstraw, Catchweed—*Galium aparine* L.

Catchweed bedstraw seeds

Catchweed bedstraw leaves in whorls

Catchweed bedstraw fruit and stem

Seed

- Size: 1.9—3.0 mm long, 1.6—2.5 mm wide, thickness 1.4—2.3 mm (excluding bristles)
- Shape and texture: circular to kidney-shaped; seed scar is depressed, one end has the shape of a bowling pin; surface covered with stiff, hooked spines, surface between the bristles is rough
- Color: generally gray-brown; the surface is dark-brown and the bristles are transparent, white or orange flecked

Catchweed bedstraw plant

Leaf shape and arrangement: Leaves—Leaves are narrow, simple, and linear, 1—8 cm long and 2—3 mm wide, stiff hairs, mostly in whorls of six to eight arrangements around the stem; leaves and stems are very rough to the touch due to the presence of very short, stiff hairs making the plant attach to clothing and animal fur.

Stolon/Rhizome/Roots: No stolons or rhizomes; stems are weak, square in cross-section.

Inflorescence: Flowers are minute, white and borne on short branches in singles or in clusters of two to three in the leaf axils each having four petals, flowers 2—3 mm wide.

Bedstraw, catchweed—*Galium aparine* L.

Family: Rubiaceae
Other common names: Cleavers, bedstraw, catchweed, goose-grass, scratch-grass, gripgrass
Life cycle: Summer or winter annual
Native to: North America

Distribution and Adaptation
- Wide spread throughout North America.
- Thrives in shady areas, often on the north side of buildings; often found in gardens, waste places, moist woods, meadows, pastures, fencerows, and in cultivated fields.

Morphology/Growth Pattern
- Often a tangling, mat-forming, shallow-rooted plant that may grow up to 6.6 ft (2 m) long.
- Dispersed typically by animals and humans because of the hooked hairs on the seeds.
- Reproduces by seed.

Use and Potential Problems
- A serious problem in cultivated fields, particularly in hay and grain.
- Can interfere with crop growth and reduce seed yield. In addition, bedstraw seed is difficult to separate in seed cleaning operations thereby reducing quality of harvested seed. For these reasons, effective herbicides for control of bedstraw in grass seed crops are desired.

Toxicity
- None recorded.

Similar Species
- Similar plant to catchweed bedstraw includes *smooth bedstraw* (*Galium mollugo* L.). However, smooth bedstraw unlike catchweed bedstraw is a perennial that spreads by rhizomes and stolons.

Reference
- Catchweed bedstraw [18,24,36,39,44,46,66,72]

Beggarticks, Devils—*Bidens frondosa* L.

Devil's beggarticks leaf arrangements *Source: Wise Co.*

Devil's beggarticks plant *Source: Wise Co.*

Devil's beggarticks flower head *Source: Wise Co.*

Leaf shape and arrangement—Leaves: Opposite, pinnately dissected to three to five divisions; leaf margins are toothed.

Stolon/Rhizome/Roots: No stolons or rhizomes; stems erect, smooth to slightly hairy; shallow, branched taproot with fibrous root system.

Inflorescence: Flowers are surrounded by 5—10 green bracts, which are longer than the orange-yellow petals; disk flowers in the center of the flower are brownish yellow; some heads may have only disk flowers.

Beggarticks, devils—*Bidens frondosa* L.

Family: Asteraceae
Other common names: Sticktight, Devil's pitchfork, bur marigold, devil's boot-jack, pitchfork weed, preacher's lice, beggarticks, tickseed sunflower
Life cycle: Summer annual
Native to: North America

Distribution and Adaptation
- Found throughout the southeast region but most common in the eastern and north-central states.
- Found in cultivated fields, ditches, fence-rows and roadsides; often appears in low moist fields.

Morphology/Growth Pattern
- The plant is erect and grows up to 1–4 ft (0.30–1.2 m) high, the stem is branched, smooth and often purplish; leaves are opposite, smooth 2–4 in. (5–10 cm) long and divided into three to five leaflets; flowers are yellow to orange in color, flat, rounded, and compact.
- Reproduces by seed; seed sticks to clothing, wool and any objects and transports from areas.

Use and Potential Problems
- No commercial value, primarily weeds of landscapes, nurseries, but also found in pastures, roadsides.

Toxicity
- None recorded.

Similar Species
- *Nodding beggarticks* (*Bidens cernua* L.) is similar to Devil's beggarticks. However, nodding beggarticks leaves are not petioled, are undivided, nodding and the fruits generally have three to four horns, while Devil's beggarticks leaves are long stocked (petioled) and are divided into three to five sharply toothed leaflets, do not have nodding heads.

Reference
- Devils beggarticks [17,24,32,33,46,48,61,67,71,73,75]

Bergamot, Wild—*Monarda fistulosa* L.

Wild bergamot stem and leaves

Wild bergamot plant *Source: John Wright*

Wild bergamot flower *Source: John Wright*

Leaf shape and arrangement—Leaves: Oval- to lance-shaped and sharply serrated/toothed; leaves arranged on stem opposite from each other; 5−8 cm long, 2−4 cm wide; gray-green in color.

Stolon/Rhizome/Roots: No stolons, slender creeping rhizomes; square, smooth stems, few branches; glabrous or pubescent at the nodes and on the angles.

Inflorescence: Flower clusters are solitary at the ends of a square stem; 20−50 flowers appear in a dense, rounded cluster about 3.5 cm long; lavender-pink tubular or rarely white; leafy bracts under the flower heads are often tinted with lavender.

Bergamot, wild—*Monarda fistulosa* L.

Family: Lamiaceae
Other common names: Bee-balm, Monarde fistuleuse, horsemint
Life cycle: Cool-season perennial
Native to: North America

Distribution and Adaptation
- Found across Canada and the United States (except Florida).
- Found in wooded slopes and meadows, moist to medium dry valleys usually occurring in small isolated clusters.
- Common in calcareous regions; prefers moist, well-drained soils.

Morphology/Growth Pattern
- Like other members of the mint family, this showy plant with aromatic leaves has square stems and opposite leaves.
- The plant can grow up to 2.3−6 ft (0.7−1.8 m) tall.
- Rhizomatous with a tall, slender, erect stem, simple or occasionally branching and slightly hairy; leaves with petioles and broadly rounded at the base and tapering at the tip; leaf margins toothed; flowers in terminal head subtended by a number of greenish or pinkish bracts.
- Reproduces by seed and rhizomes.

Use and Potential Problems
- Generally, this plant is considered a weed in pastures and rangeland environments.
- Not a good feed source for cattle or game and is somehow bitter to goats and sheep.
- Flowers are very attractive to bees and hummingbirds.

Toxicity
- None recorded.

Similar Species
- Similar plants include *white bergamot* (*Monarda clinopodia* L.); however the flowers of white bergamot are very pale and the bracts underneath the flower are white; both plants have similar fragrance.

Reference
- Wild bergamot [34,38,39,46−48,58,75]

Bindweed, Field—*Convolvulus arvensis* L.; Bindweed, Hedge—*Calystegia sepium* (L.) R. Br.

Field bindweed seeds

Hedge bindweed plant *Source: Rocky Lemus*

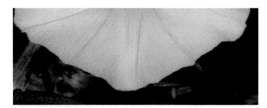

Hedge bind weed flower

Field bindweed flower *Source: Lachlan Cranswick*

Seed
- Size: 3—5 mm long, thickness 0.7—3.0 mm
- Shape and texture: three-angled (orange-slice shaped), ovoid, flat on one or two sides, the other side rounded; surface uniformly covered with small irregular tubercles, very dull, coarsely roughened
- Color: dark gray-brown colored
- Seeds for hedge bindweed (not shown) are twice as large as field bindweed, much darker and smoother, and with light color depression

Field bindweed plant *Source: Wise Co. VA*

Leaf shape and arrangement—Leaves: Field bindweed leaves are spade-shaped and are parallel to the stem, glabrous, alternate, simple long petiole with acute basal lobes; 2—6 cm long and about 3 cm wide; hedge bindweed leaves are triangular, with two distinct angles at each basal point (square lobes at bottom of leaf resembling dog ears).

Stolon/Rhizome/Roots: No stolons, both field and hedge bindweeds are rhizomatous; stems are slender, prostrate, or twining, may be with or without hairs; field bindweed has extensive and deep rhizomes and root system which may penetrate to a depth of 2.5—6 m; hedge bindweed has extensive but shallow root system with extensive rhizomes.

Inflorescence: Field bindweed flowers are white to pink in color, solitary or rarely two on axillary peduncles (one-flowered); has two small scale-like bracts on the peduncle (flower stem) from 1.3 to 5 cm below the flower. Hedge bindweed has large white flowers, the base of each flower is enclosed by two large green bracts; hedge bindweed *does not have bracts (leaf-like appendages) between bloom and stem.*

Bindweed, field—*Convolvulus arvensis* L.

Family: Convolvulaceae
Other common names: Wild morningglory, small-flowered morningglory, European bindweed, corn-bind, bear-bind, creeping Jenny, green-vine
Life cycle: Cool-season perennial
Native to: Europe

Distribution and Adaptation
- Distributed throughout the northern United States and southern Canada; most abundant and troublesome in western states.
- Found in grasslands, open forests, fields, roadsides, waste places, and disturbed areas.

Morphology/Growth Pattern
- This perennial plant can grow up to 23 ft (7 m) long.
- Trailing along the ground or climbing; the creeping white rhizome can grow up to 210 ft (30 m) in length and 30 ft (9 m) deep.
- Few to many stems clustered from branched crown, spreading outward and often twinning stems.
- Reproduces by seed and rhizomes.

Use and Potential Problems
- Field bindweed in many states is a prohibited noxious weed.
- Depending on the extent of invasion, it can cause 30%−100% crop loss.

Toxicity
- Digestive disturbance toxins and calystegine (alkaloids).

Similar Species
- Field bind weed is often confused with *hedge bindweed* (*Calystegia sepium* (L.) R. Br.), but hedge bind weed has much larger and shiny green leaves and a much larger white flower. The leaf bases of field bindweed are pointed or rounded. Hedge bindweed leaf bases are squared. Field bindweed also can be easily identified by a pair of small bracts located near to well above the middle of the peduncle.

Reference
- Field bindweed [1,12,24,25,39,48,65,72]

Bittercress, Hairy—*Cardamine hirsuta* L.

Pinnate leaves with one to three pair of alternate leaflets

Bittercress plant *Source: James Altland*

Bittercress flower and seedhead

Leaf shape and arrangement—Leaves: Margins shallowly toothed or with a few lobes; 3−8 cm long, 0.7−2.5 cm wide, pinnate with one to three pairs of alternate rounded to wedge-shaped leaflets, hairy near the leaf base; basal leaves with or without hair more numerous than the much smaller and fewer stem leaves.

Stolon/Rhizome/Roots: No stolons or rhizomes; flowering stems are smooth, erect, usually branched at the base with few leaves; much branched taproot.

Inflorescence: Flowers white, in dense cluster at the end of stems; four petals; fruits in flattened capsule, 10 times longer than broad.

Bittercress, hairy—*Cardamine hirsuta* L.

Family: Cruciferae
Other common names: Hoary bittercress, common bittercress, popping cress, Pennsylvania bittercress, small bittercress
Life cycle: Winter or summer annual
Native to: Europe

Distribution and Adaptation
- Most common in the southeastern United States.
- It grows on moist, sandy, or clay soils in cultivated and waste places.

Morphology/Growth Pattern
- A winter or summer annual, less frequently biennial and grows up to 3–9 in. (8–23 cm) tall.
- Stems arise from a basal rosette of dark green, dissected leaves; ascending or erect stem.
- Reproduces by seed.

Use and Potential Problems
- Weed of turfgrass, landscapes, nurseries, greenhouses, and roadsides.

Toxicity
- None recorded.

Similar Species
- Small flowered *bittercress* (*Cardamine parviflora* L.) is similar to hairy bittercress; however, small flowered bittercress lacks the minute white hairs at the base of the leaves and unlike hairy bittercress the margins of the leaves are lobed; also the mature plant of small flowered bittercress does not have basal leaves.

Reference
- Hairy bittercress [24,44,47,58,72,75]

Blueweed or Viper's Bugloss—*Echium vulgare* L.

Blueweed seeds

Blueweed stem and leaves

Blueweed flowers

Seed
- Shape: mericarps ovoid-oblong
- Color: tan

Blueweed plant

Leaf shape and arrangement—Leaves: Rosette leaves narrow to a short petiole, leaves are oblong to linear-lanceolate in outline, 5–15 cm long and reaching 3 cm in width; the leaves on the flowering stem progressively become smaller toward the top of the plant; leaves have white "speckles" that give the leaves a dimpled appearance and also have relatively long white hairs.

Stolon/Rhizome/Roots: No stolons or rhizomes; stems erect, branching, reaching 0.76 m in height; stems with long bristly hairs.

Inflorescence: Showy, tubular blue flowers on curled branches, each with protruding red stamens in one-sided clusters on lateral branches; clusters uncoil as flowers bloom; flowers 2 cm long.

Blueweed or Viper's bugloss—*Echium vulgare* L.

Family: Boraginaceae
Other common names: Viper's bugloss, blue-weed, blue devil, common Viper's-bugloss, common Vipersbugloss
Life cycle: Cool-season biennial
Native to: Europe

Distribution and Adaptation
- Can be found throughout the United States except the southeast and the southwest.
- Found in fields, roadsides, noncrop areas and waste places.
- Requires well-drained soils and can grow in low fertility soils.
- The plant can grow well on acidic or neutral to basic soils; cannot tolerate shade.

Morphology/Growth Pattern
- This biennial has a rosette growth habit during the first year of growth and produces flowering stems the second year; grows 2−7 ft (0.6−2 m) tall.
- Is a taprooted plant and develops a rosette of large coarse leaves covered with stiff hairs with swollen, reddish, or black bases; the flowering stem is also roughened by stiff hairs with enlarged bases; the stem leaves are sessile and have similar hairs to the rest of the plant; leaves have dent or "dimple" appearance; flowers are numerous and arranged in dense one-sided racemes; flowers are blue, irregular in shape.
- Have both the female and male organs on the same plant.

Use and Potential Problems
- No value as livestock feed, considered a weed.
- Leaves are poisonous.

Toxicity
- None recorded.

Similar Species
- At the rosette stage, this plant can be confused with *curly dock* (*Rumex crispus* L.); however, curly dock usually has curled leaves at the base and does not have white speckles on the stem or dents on the leaves like blueweed.

Reference
- Blueweed [47,48,58,61,75]

Buckwheat, Wild—*Fallopia convolvulus* (L.) Á Löve

Wild buckwheat seeds

Wild buckwheat leaf *Source: James Altland*

Wild buckwheat seedhead *Source: James Altland*

Seed
- Size: 3.5—4.0 mm long, 2.4—2.7 mm wide
- Shape and texture: three-sided and three-angled, broadest at the middle with blunt base and narrow apex, more sharply triangular than field bindweed; surface smooth, dull but the angles are glossy; under high magnification the surface is finely striate
- Color: black undercoat under the light brown hull

Wild buckwheat field *Source: James Altland*

Leaf shape and arrangement—Leaves: Heart-shaped leaves, 2—6 cm long, with basal lobes and a small papery sheath that encircles the stem at the leaf base (*ocreas at petiole bases*); leaf gradually tapers to form a point at the tip, hairless, giving the plant a glossy texture.

Stolon/Rhizome/Roots: No stolons or rhizomes; erect stem at first, then becomes twinning or creeping and branched at the base, fibrous root systems.

Inflorescence: Clusters of inconspicuous green flowers form in leaf axils or at the end of stems; fruit is a three-angled achene that is black, 3—4 mm long, and enclosed in the green, winged sepals.

Buckwheat, wild—*Fallopia convolvulus* (L.) Á Löve

Family: Polygonaceae
Other common names: Black bindweed, knot bindweed, bear-bindweed, ivy bindweed, climbing bindweed, climbing buckwheat, corn-bind
Life cycle: Summer annual
Native to: Europe

Distribution and Adaptation
- Found throughout the United States and northern Canada.

Morphology/Growth Pattern
- Wild buckwheat is an annual with smooth, slender, twinning, or creeping stems; the plant branches at the base; leaves are heart-shaped and pointed with smooth edges, alternate on the stem; grows up to 6.5 ft (2 m) in length.
- Reproduces by seed.

Use and Potential Problems
- Weed of landscapes, pastures, nurseries, vegetables, and agronomic crops especially grains.

Toxicity
- Photosensitization agents.

Similar Species
- Wild buckwheat is similar in appearance to morningglories, except it has ocreas at the base of the leaf petioles, also often confused with *field bindweed* (*Convolvulus arvensis* L.) because of the vine growth habit and similar leaf shape. However, buckwheat is an annual that grows from seed, while field bindweed is a perennial that often regrows from roots; also unlike bindweed, the leaves of wild buckwheat are heart-shaped and the flowers are minute.

Reference
- Wild buckwheat [1,12,18,48,61,66,72,73]

Bur Cucumber—*Sicyos angulatus* L.

Bur cucumber plant

Leaf shape and arrangement—Leaves: Alternate, round to broadly heart-shaped with five pointed shallowly angled lobes, toothed margins; long petioled; sticky-hairy, 6—20 cm long and 6—20 cm wide.

Stolons/Rhizomes/Roots: No stolons or rhizomes; the stems longitudinally ridged, sticky-hairy especially at leaf nodes, with branched tendrils from the side of the leaves; fibrous root system.

Inflorescence: Flowers are greenish white with five sepals and five petals; cluster of fruits, oval, three to five berries each containing one seed, fruits covered with long stiff bristles and short hairs.

Wild cucumber plant *Source: Nathan O'Berry*

Bur cucumber fruits *Source: Fred Fishel*

Wild cucumber leaves and flowers *Source: Nathan O'Berry*

Bur cucumber—*Sicyos angulatus* L.

Family: Cucurbitaceae
Other common names: One-seeded bur cucumber, star cucumber, nimble kate
Life cycle: Summer annual
Native to: North America

Distribution and Adaptation
- Found throughout the northeastern United States.
- Plant can grow well on moist, dump rich soils.

Morphology/Growth Pattern
- This vining plant is easily identified by its branched tendrils opposite each leaf, big leaves, clustered flowers, and spiny cluster of fruits that resemble cucumbers. The plant grows very fast, producing shoots up to 26 ft (8 m) long in the first year from seed.
- Reproduces by seed.

Use and Potential Problems
- Primarily a weed that grows along fence rows, along creeks and waste places. However, in recent years, the plant has been found on agricultural lands where no-till farming planting is commonly practice. Can be a major problem in mechanically harvested crops or vegetables where the plant interferes with harvesting equipments.

Toxicity
- None recorded.

Similar Species
- Plants that resemble bur cucumber include *wild cucumber* (*Echinocystis lobata* L.). However, wild cucumber is distinguished from bur cucumber by its more deeply lobed leaves that are heart-shaped or star-shaped, stems almost glabrous (hairless), and flowers with six sepals and six petals. In addition, bur cucumber has clusters of 3—10 spiny fruits, whereas wild cucumber has 1 fruit at each node.

Reference
- Bur cucumber [17,24,36,61,66,72,75]

Burdock, Common—*Arctium minus* (Hill) Bernh.

Common burdock seeds

Seed
- Size: 5.7–6.1 mm long, 2.2–2.5 mm wide, thickness 0.8–1.0 mm; seeds originate in a bur covered in hooked
- Shape and texture: wedge-shaped, and curved on one or both sides; similar to wild sunflower, but curved at end and not bulged in center like wild sunflower; surface finely textured, narrow central ridge on each broad surface, under high magnification can see many small lengthwise wrinkles
- Color: grayish to dark brown mottled with black

Common burdock plant

Leaf shape and arrangement—Leaves: The first year, rosette leaves with long, thick, and hollow petioles, leaves are large, 50 cm long and 40 cm wide, heart-shaped, dark green, smooth above, whitish green and woolly hairy beneath, wavy and toothed margins, like rhubarb except pubescent.

Stolon/Rhizome/Roots: No stolons or rhizomes; the stem is thick, stout, grooved, rough, has multiple branches, and grows to 61–182 cm; large thick fleshy taproot, as deep as 1 m below the soil surface.

Inflorescence: Second year forms rough, hairy flowering stalk; red-violet flower heads surrounded by numerous hooked bracts that form a bur-like cup; burs similar to cocklebur but shorter and bushier; spines on the bur are not as stiff as cocklebur; spines have hooks on the end.

Common burdock flowers

Burdock, common—*Arctium minus* (Hill) Bernh.

Family: Compositae
Other common names: Lappa minor, wild rhubarb, great burdock, beggar's button, smaller burdock, clotbur, cockoo-button, cockle-button, lesser burdock
Life cycle: Cool-season biennial
Native to: Europe

Distribution and Adaptation
- Found throughout North America.
- Found in waste places, feed lots, wooded areas, roadsides, abandoned farm lands, and fence rows.

Morphology/Growth Pattern
- Produces a basal rosette during its first year and the rosette could reach up to 3.3 ft (1 m) wide, the second year of growth produces flowering stem which grows up to 7 ft (2 m) tall; the stem is hollow and grooved and has fleshy taproot.
- The burs attach themselves to clothing, hair and wool.
- Reproduces by seed.

Use and Potential Problems
- Common burdock is a weed of landscapes, nurseries, and pastures; not a problem in cultivated land.
- The value of wool is significantly reduced when the dry burs are attached to the wool. Cows that have consumed a large quantity of burdock have been reported to produce milk with bitter taste.
- Burs cause local irritation and possible intestinal hairballs; all animals, particularly livestock, are affected.
- Common burdock roots are consumed as a vegetable in Japan.

Toxicity
- Potential diuretic effects and allergic reactions when hooked bristles contact skin.

Similar Species
- Burdock is often confused with *common cocklebur* (*Xanthium strumarium* L.) (a far more dangerous plant). Burdock burs are rounder and have softer, more velcro-like hairs than cocklebur. Cocklebur burs are oblong and have hooked spines on the bur, and have, on the end of the bur, two spines which are larger and not as strongly hooked

Reference
- Common burdock [12,17,48,65−67,72,73]

Buttercup, Bulbous—*Ranunculus bulbosus* L.;
Buttercup, Corn—*Ranunculus arvensis* L.;
Buttercup, Tall Field—*Ranunculus acris* L.

Bulbous buttercup seeds

Bulbous buttercup leaf *Source: Rocky Lemus*

Bulbous buttercup flower *Source: NCSU*

Seed

- Size: 3—3.5 mm long, 2.3—2.8 mm wide (bulbous buttercup); tall buttercup seeds are 2.2—3.3 mm long, 2—2.6 mm wide
- Shape and texture: both buttercup seeds are broadly egg-shaped, seed has a hook-like protrusion; surface smooth
- Color: seeds of bulbous buttercup are reddish-brown with a yellow margin; tall buttercup light brown to dark brown

Bulbous buttercup plant *Source: NCSU*

Leaf shape and arrangement—Leaves: The basal leaves are long, hairy petioles, three-parted with each division very lobed or dissected; the terminal segment is stalked and the two laterals sessile (attached directly to the main leaf petiole); the stem leaves are smaller, sessile (no petiole) and are less lobed.

Stolon/Rhizome/Roots: No stolons or rhizomes; has a basal corm (bulb) from which the stems arise; stems are erect, branching, somewhat round, white and hairy; fibrous root systems.

Inflorescence: Single flowers on flower stalk at the end of the stems; flowers bright, shiny having five to seven yellow petals; flowers are broadly rounded at the tip and have wedge-shaped bases; five green sepals sharply turned-down (curved back) toward the stem.

Buttercup, bulbous—*Ranunculus bulbosus* L.

Buttercup, corn—*Ranunculus arvensis* L.

Family: Is a member of crowfoot family (Ranunculaceae)
Other common names: St. Anthony's turnip, bulbous crowfoot, yellow weed, blister flower, gowan, frogsfoot, goldcup
Life cycle: Cool-season perennial
Native to: Europe

Distribution and Adaptation
- Common in the northeastern, southeastern, and western United States and in Canada.
- Grows in wet soils and in marsh areas; pastures heavily infested with buttercups are usually characterized by acid soils and poor fertilization.

Morphology/Growth Pattern
- A tufted-perennial, forms a basal rosette of three-parted leaves; stems rise from a bulb-like swelling at the base (corm) of the stem; the bulb-like structure resembles turnip hence being referred to as St. Anthony's turnip; can grow up to 8–24 in. (20–61 cm).
- Stems of bulbous and tall field buttercups are erect but can grow more prostrate in turfgrass and lawns.
- Plant over winters as corm but reproduces by seed.

Use and Potential Problems
- Common weeds of turfgrass, lawns, pastures, hay fields, and landscapes; bulbous buttercup is a rare weed in cultivated crops.
- Buttercups are unpalatable and animals would not consume buttercups unless other desirable feeds are scarce.

Toxicity
- Ranunculin.

Similar Species
- Several buttercup species are found in the southern United States; *tall field buttercup* is similar to bulbous buttercup but the leaves of tall field buttercup are not deeply lobed, and does not have a bulb-like stem base and the sepals of the flower head on tall field buttercup does not curve back toward the stem; creeping buttercup is also similar to the two buttercups described above but like tall field buttercup lacks a bulb-like stem base, the middle lobe of the leaf has a long stock and the three lobes are not as deeply toothed as bulbous or tall field buttercups, additionally, creeping buttercup has larger flowers (than the other two buttercups) and the sepal is not curved back toward the stem like bulbous buttercup; creeping buttercup has stolons.

Reference
- Bulbous buttercup [12,18,24,25,33,47,48,61,72,75]

Butterflyweed—*Asclepias tuberosa* L.

Butterflyweed seed

Seed
- Shape and texture: flat, much wider one end
- Color: light brown
- Seed per pound: 87,000

Butterfly weed plant

Butterflyweed opposite leaf

Leaf shape and arrangement—Leaves: Numerous, alternate on the stem, many in number and lance-shaped, rather rough leaves from 5 to 15 cm long.

Stolon/Rhizome/Roots: No stolons or rhizomes; stems stout, hairy, solitary and often clustered from a thick root crown, branching from the upper nodes; produces a very deep, large, branching, white, and fleshy taproot.

Inflorescence: The clustered, compacted, bright, oddly shaped, orange flowers are borne at the ends of the branching stems.

Butterflyweed flowers

Butterflyweed—*Asclepias tuberosa* L.

Family: Asclepiadaceae
Other common names: Butterfly milkweed, pleurisy-root, chigger-weed, Canada-root, Indian posy, orange-root, orange swallowwort, tuberroot, whiteroot, windroot, yellow or orange milkweed.
Life cycle: Warm-season perennial
Native to: North America

Distribution and Adaptation
- Found throughout the eastern half of the United States; but found in greatest abundance in the south.
- Extremely hardy and long-lived perennial.
- Requires a well-drained sandy or gravelly soil in full sun; flourishes in open or in pine woods usually along the banks of streams.

Morphology/Growth Pattern
- Tall, stout forb with a deep rooted system; the stems are coarsely hairy, erect, 1–2 ft (0.30–60 m) tall; usually appear one to five in a group and contain very little milky sap; leaves alternate and attached on an erect hairy stem; leaves are narrow and are lance-shaped; flower heads are composed of many individual orange, bright, beautiful flowers.
- Reproduces by seed and rootstocks; reproduction by seed might take longer than a year for full production.

Use and Potential Problems
- Mainly considered as a "beautiful weed."
- Animals do not like butterflyweed and if they consume the plant it is not by choice.
- If animals are forced to eat this plant in a large quantity, it may produce toxic effects.
- The flowers shaped like hourglasses and having a special nectar holder, attract many insects while in bloom.
- The flowers produce a large quantity of nectar which attract butterflies throughout the growing season.

Toxicity
- Cardiotoxins.

Similar Species
- Similar species include *common milkweed* (*Asclepias syriaca* L.) and *swamp milkweed* (*Asclepias incamata* L.); unlike butterflyweed both these species produce milky sap. Common milkweed has pinkish-purple flowers, while swamp milkweed produces flesh or rose colored flowers.

Reference
- Butterflyweed [12,33,34,38,39,48,57,58]

Campion, White—*Silene latifolia* Poir. subsp. *alba* (Mill.) Greuter & Burdet

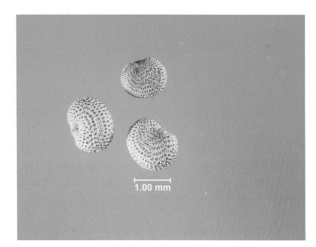

White campion seed

White campion leaves arranged opposite on the stem *Source: Lachlan Cranswick*

White campion flower with five petals *Source: Lachlan Cranswick*

Seed (White Campion)
- Size: about 1.5 mm long smaller than corn cockle
- Shape and texture: nearly round but has a little nick
- Color: covered with rows of warty bumps; gray and dull

White campion plant *Source: Lachlan Cranswick*

Leaf shape and arrangement—Leaves: Opposite, oblong, 2−10 cm long and 2 cm wide, narrow pointed at the tip, lower leaves are stalked (with petiole), while the upper leaves are sessile; leaves are light green in color; entire plant surface is hairy.

Stolon/Rhizome/Roots: No stolons or rhizomes; stems covered with fine short hairs; thick, tough branching taproot.

Inflorescence: White to tan five-petaled flowers that are split or lobed shaped like "v" flowers in characteristic bladder-like case which are fused inflated sepals with 10 (male flowers) and 20 (female flowers); shorter sepals than corn cockle; multiple flower per peduncles; *fruit* ovate capsule, 10−15 mm long; each female plant can produce over 25,000 seeds.

Campion, white—*Silene latifolia* Poir. subsp. *alba* (Mill.) Greuter & Burdet

Family: Aryophyllaceae
Other common names: White cockle, evening campion evening lychnis, snake cuckoo, thunder flower, bull rattle, white robin
Life cycle: Cool-season, annual, biennial or short lived perennial
Native to: Europe

Distribution and Adaptation
- Found in the northern half of the United States and Canada.
- Frequently found at low to high elevation and moist disturbed habitats but sometimes in undisturbed meadows and often forest.

Morphology/Growth Pattern
- Leaves are opposite, connected at the base with ridge; the basal and lower leaves are petioled; plant tall simple or branched from the base, erect or low growing with a stout often laterally branched taproot; plant can grow up to 1−2.5 ft (0.30−0.76 m) high; numerous white funnel shaped flowers; male and female plants are produced in separate plants; seedpods are swollen, hairy and each pod has 10 short teeth at the top.
- Flower appears in the evening and closes in the morning hence called evening campion.
- Reproduces by seed.

Use and Potential Problem
- Problematic weed in cultivated crops and pastures; also is a host to some viruses.

Toxicity
- None recorded.

Similar Species
- White campion also known as white cockle also commonly confused with *corn cockle* (*Agrostemmas githago*). Corn cockle sepal is slender, well beyond the petal alike white campion; *bladder campion* (*Silene vulgaris* L.) similar in appearance to white campion, however, bladder campion does not have hairs on the leaves like white campion.

Reference
- White campion [6,19,36,38,39,47,61,72,74]

Carpetweed—*Mollugo verticillata* L.

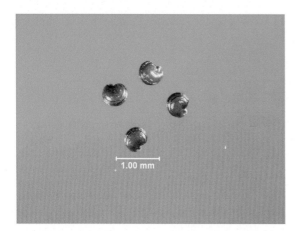

Carpetweed seeds

Seed
- Size: 0.5–0.6 mm long, about 0.4 mm wide, thickness approximately 0.4 mm
- Shape and texture: kidney-shaped, flattened, thick toward the outer edge, coiled, projection on indented margin; surface ridged, shiny
- Color: orange brown, the ridge and the small tooth are darker, under high magnification the ridge appear translucent

Carpetweed plant *Source: NCSU*

Carpetweed leaves, stem, and flowers (NCSU) *Source: NCSU*

Leaf shape and arrangement—Leaves: Sessile, pinnately veined, alternate but leaves appear whorled with three to eight at each stem node; leaves 1–3 cm long, widest above the middle and tapering to the base (spoon-shaped); are long, narrow, or linear shaped; leaves can reach 2.5 cm in length; the stipule resembles the leaves; light green in color.

Stolon/Rhizome/Roots: No stolons or rhizomes; stem smooth, low growing, radially spreading from a central root, no rooting at the nodes; and much branched; green in color and are often 5–30 cm long; have a small branched taproot.

Inflorescence: Flowers are sessile, white, small (4–5 mm wide) arranged in clusters of two to five on slender stalks arising from leaf axils, sepals green with white margins.

Carpetweed flower

Carpetweed—*Mollugo verticillata* L.

Family: A member of the carpetweed family (Aizoaceae)
Other common names: Indian chickweed, green carpetweed
Life cycle: Summer annual
Native to: North America

Distribution and Adaptation
- Found throughout temperate North America; in the United States, commonly found in the eastern part of the country.
- Adapted to rich soils also commonly found on dry, gravelly, or sandy soils.

Morphology/Growth Pattern
- Much branching yellowish green stems, forming prostrate mats on the soil surface; grows up to 12 in. (30 cm) long, short taproot with very few branches.
- Reproduces by seed.

Use and Potential Problems
- Weed in fields, gardens, waste places, roadsides, cultivated agronomic and horticultural crops, and in new or thin turfgrass fields.

Toxicity
- None recorded.

Similar Species
- Carpetweed resembles bedstraw (*Galium aparine* L.). Although bedstraw has whorls of leaves like carpetweed, bedstraw differs from carpetweed by its square stems. Bedstraw leaves and stems alike carpetweed have very short, stiff hairs.

Reference
- Carpetweed [14,19,24,44,58,66,72]

Carrot, Wild—*Daucus carota* L. subsp. *carota*

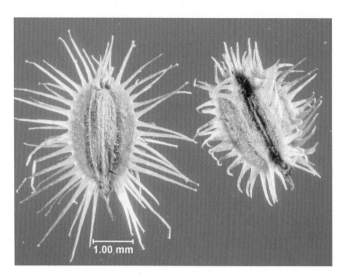

Wild carrot seeds

Wild carrot leaves

Wild carrot flower head *Source: Peter Sforza*

Seed
- Size: 3.2–4.2 mm long and 1.7–2.6 mm wide
- Shape and texture: elliptical in shape with flat side, it resembles a bug with no legs; the seed surface has four rows of short curved spine and five length-wise rows of ridges from one end to the other
- Color: yellowish to grayish

Wild carrot plant *Source: Peter Sforza*

Leaf shape and arrangement—Leaves: Definite "carrot" leaves, leaves dissected into fine divisions (lacy), hairy (similar to cultivated carrot), stem leaves are sessile with a sheathing base, basal leaves long petioled.

Stolon/Rhizome/Roots: No stolons or rhizomes; the taproot is less than 5 cm in diameter, white and woody.

Inflorescence: Small white flowers composed of five small green sepals, five white petals, five stamens, and two styles; below each umbel is a whorl of finely divided green bracts with three to five branches; after flowering the umbel stalk turn upward closing the umbel and forming "bird's nest"; fruit breaks into 2 seeds, 2–4 mm long; each plant produces up to 40,000 seeds.

Carrot, wild—*Daucus carota* L. subsp. *carota*

Family: Apiaceae
Other common names: Queen Anne's lace, bird's nest, devil's-plague, lace-flower
Life cycle: Cool-season, biennial
Native to: Europe southwest Asia and north Africa

Distribution and Adaptation
- Found throughout the United States except the north central plains and throughout Canada.
- Can be frequently found in dry areas and rocky soils and in soils that are slightly or moderately acidic.
- Originally introduced as an ornamental.

Morphology/Growth Pattern
- The taprooted biennial herb can grow 2–4 ft (0.60–1.2 m) tall; the first year the plant grow as a rosette with a deep taproot; stems are erect, hollow, stiff-haired and sometimes branched; alternate leaves two to three times pinnately compound, lower leaves largest and are on loner petiole compared with those leaves located on the middle and upper part of the stem; white flowers borne in umbrella-shaped clusters.
- Wild carrot can be distinguished from cultivated carrot by its white taproot.
- The root and the foliage has a strong carrot-like odor.
- Reproduces by seed.

Use and Potential Problems
- Weed of old pastures, meadows, roadsides, and waste areas; can be controlled by cultivation.

Toxicity
- Skin irritation; if large amounts consumed by cow, can cause off flavor in milk.

Similar Species
Similar species include *pineappleweed—Matricaria matricarioides* (Less.) C.L. Porter, *common ragweed—Ambrosia artemisiifolia* L., and *common yarrow—Achillea millefolium* L. All these species can be distinguished from each other by the serration and shape of the leaves and by hairs on stems and leaves.

Reference
- Wild carrot [6,12,14,19,24, 25,33,36,39,47,61,67,72,74,76]

Chickweed, Common—*Stellaria media* (L.) Vill.; Chickweed, Mouse-Ear—*Cersatium fontanum* Baumg. subsp. *vulgare* (Hartm.) Greuter & Burdet

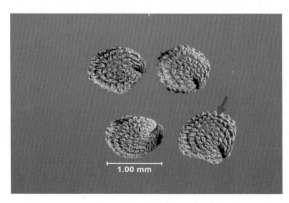

Common chickweed seeds; roughened by curved rows of minute tubercles

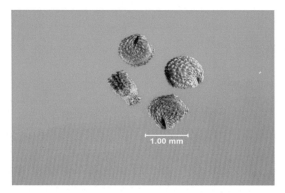

Mouse-ear chickweed seeds

Seed

Common chickweed

- Size: 0.8−0.9 mm long, 0.8−0.9 mm wide
- Shape and texture: irregularly circular with base slightly elongated, two flat faces that join the margin at nearly right angles, seed appear folded; surface tuberculate, dull, the tubercles are arranged in rows around marginal face and in concentric circle
- Color: straw to pale reddish brown

Mouse-ear chickweed

- Size: 0.6−0.7 mm long, 0.6−0.7 mm wide, thickness 0.3−0.5 mm
- Shape and texture: irregular with some rounded and some angular edges; the round hilum on one corner has a small notch formed by incurved tissue; many papillae are elongated, under magnification papillae are shiny and transparent
- Color: light orange brown

Common chickweed plant *Source: Lachlan Cranswick*

Mouse-ear chickweed flower *Source: Lachlan Cranswick*

Chickweed, common—*Stellaria media* (L.) Vill.

Chickweed, mouse-ear—*Cersatium fontanum* Baumg. subsp. *vulgare* (Hartm.) Greuter & Burdet

Mouse-ear chickweed plant *Source: NCSU*

Common chickweed leaves opposite from each other *Source: NCSU*

Mouse-ear chickweed *Source: NCSU*

Leaf shape and arrangement—Leaves: Of common chickweed is small, smooth (no hair), opposite, oval in shape, 1−3 cm long and 3−15 mm wide looks like a mouse's ear; lower leaves are stalked, but the upper are sessile (2−7 cm long and 1−15 mm wide). Common chickweed and mouse-ear chickweed are very similar, however, mouse-ear chickweed is densely hairy.

Stolon/Rhizome/Roots: No stolons or rhizomes; fibrous root systems; stems are branched and have rows of vertical fine hairs on the surface (common chickweed); roots at the stem nodes (mouse-ear chickweed).

Inflorescence: Flowers usually in clusters of three at the end of stems; small white flowers with 5 deeply notched petals giving the appearance of 10 petals.

Common chickweed leaves and flowers *Source: NCSU*

Chickweed, common—*Stellaria media* (L.) Vill.

Family: Caryophyllaceae
Other common names: Starwort, satin flower, starweed, winter-weed, tongue-grass
Life cycle: Winter annual

Chickweed, mouse-ear—*Cersatium fontanum* Baumg. subsp. *vulgare* (Hartm.) Greuter & Burdet

Other common names: Small chickweed, spring mouse-ear, sticky chickweed
Life cycle: Cool-season perennial
Native to: Europe

Distribution and Adaptation
- Found throughout the United States.
- Commonly found in gardens, lawns, hay meadows, and pastures.
- Does not thrive in direct sunlight, but flourishes in shady places; tolerates cool and moist environments.

Morphology/Growth Pattern
- Both chickweeds have a prostrate or creeping growth habit and are low-growing mat-formers; trailing stem up to 2.5 ft (0.76 m) long.
- Mouse-ear chickweed is a perennial weed, while common chickweed is an annual.
- Reproduces by seed.

Use and Potential Problems
- Chickens graze chickweed (where it gets its name).
- Both chickweeds are common weeds in turfgrass, nursery crops, cultivated horticultural and agronomic crops.

Toxicity
- None recorded.

Similar Species
- *Common chickweed* and *mouse-ear chickweed* resemble each other. Some of the distinguishing characteristics between these two weeds are: mouse-ear chickweed is succulent, the stem and leaves are covered with sticky hairs, the stems may have a reddish-purple color and mouse-ear chickweed roots at the stem nodes while common chickweed does not. The flowers are larger than common chickweed. The petals are not as deeply divided as common chickweed. The common chickweed petals are deeply lobed giving the appearance of 10 petals instead of 5.

Reference
- Chickweed [14,18,24,32,33,44,47,48,65,66,72,74,76]

Chicory—*Cichorium intybus* L.; Cinquefoil, Sulfur—*Potentilla recta* L. "Warrenii"

Chicory seeds; looks chopped off on one end

Seed
- Size: 2.2—3.0 mm long, 0.9—1.5 mm wide; thickness 0.7—1.4 mm
- Shape and texture: slightly curved, slender, arrowhead shaped, concave on both sides; surface smooth, many lengthwise ridges, under low magnification surface appears finely textures, under high magnification can see the fine pattern of the crosswise wrinkles
- Color :light brown or grayish with black mottling
- Seed per pound: 425,923

Chicory plant

Flowers on the stem *Source: Rocky Lemus*

Leaf shape and arrangement—Leaves: Large rosette leaves, basal leaves 5—25 cm long, resembling those of the dandelion and shepherd's-purse; basal leaves absent at flowering; stem leaves alternate, clasping, and hairy and become progressively smaller toward the tip and eventually become more bract-like with toothed margins.

Stolon/Rhizome/Roots: No stolons but may have short rhizomes; stems are erect, branched, smooth with milky sap (juice); deep, fleshy taproot, simple or branched and brown in color.

Inflorescence: Flowers are bright blue; flowers directly attached on the stem (end of branch and axils), flowers 3—4 cm across with square indented (lobed) ends; flowers can appear in singles or clustered.

Chicory flower *Source: Rocky Lemus*

Chicory—*Cichorium intybus* L.

Family: Compositae
Other common names: Coffee weed, succory, blue sailors, blue daisy
Life cycle: Cool-season perennial
Native to: North Africa, Europe, and Western Asia

Distribution and Adaptation
- Scattered throughout the United States, rare in Florida.
- Found in disturbed sites such as roadsides, pastures, hayfields, turfgrass, and waste areas.

Morphology/Growth Pattern
- Plants initially produce a basal rosette of leaves that resembles dandelion.
- Stems are branched and produced during the latter part of the growing season; stems 1−6 ft (0.30−1.8 m) tall.
- Stems and leaves produce milky sap when cut.
- Reproduces by seed.

Use and Potential Problems
Weed Chicory
- The plant is sometimes used as a vegetable crop, and the roots have long been used as a substitute for coffee.

Forage Chicory
- Used for livestock feed. Variety "Puna" developed in New Zealand, marketed in the United States.
- Potential use to "mop up" excessive nutrients.
- Where adapted, will provide spring and summer growth which can supplement the cool-season forage species.
- Grown on well-drained or moderately drained soils.
- With medium to high fertility and a pH of 5.5 or greater.
- Good seedling vigor and a relatively deep taproot which provides tolerance to drought.
- Provides both spring and summer forage growth for livestock.
- Unlike most forage crops, it is an herb rather than either a grass or a legume.
- Produces leafy growth.
- Correct grazing management is essential to maintain stand longevity and maintain forage quality.
- Spring-seeded chicory can be grazed after 80−100 days, depending on climatic conditions.
- Production optimized under rotational stocking; rest period of 25−30 days depending on climatic conditions and time of the year.
- Leave a stubble height of 1.5−2 in. (cm) after grazing or cutting.
- Yield potential over 4.5 tons of dry matter per acre.
- Excellent quality, nutritive value, highly palatable to livestock.

Toxicity
- Lactones (milk taint).

Similar Species
- At the vegetative (rosette) stage, easily confused with *Dandelion* (*Taraxacum officinale* G.H. Weber ex Wiggers). However, the leaves of dandelion generally have lobes that are opposite from one another and these lobes are generally pointing in the direction of the rosette unlike those of chicory.

Reference
- Chicory [4−6,12,14,18,25,44,65,66,72]

Cinquefoil, Sulfur—*Potentilla recta* L. "Warrenii"

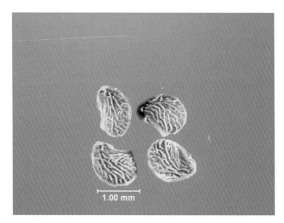

Sulfur cinquefoil seeds

Seed
- Size: 1.1–1.5 mm long, 0.7–1.1 mm wide
- Shape and texture: ear-shaped, faces slightly convex surface covered with fine, papery branched ribs
- Color: vein and margins tan; intervein spaces dark brown

Sulfur cinquefoil plant

Sulfur cinquefoil leaves and stem

Leaf shape and arrangement—Leaves: Palmately compound with five- or seven-toothed leaflets on each leaf; 2–10 cm long; the lower leaves are stalked (petioled), the upper leaves may be sessile; the upper surface of the leaves is dark green, while the undersurface is silver and hairy as in many *Potentilla* species; prominent stipule and irregularly cleft.

Stolon/Rhizome/Roots: No stolons or rhizomes; hairy erect stems with a coarse woody base and a cluster of rough, spreading dark brown roots, taprooted.

Inflorescence: Flowers flat-topped; light yellow/sulfur colored flowers with five petals, each flower producing numerous single-seeded oval achenes; the bractlets narrower than the sepals but become equal or surpasses the sepals, sepals 4–6 mm long; petals light yellow and 5–11 mm long and indented at the tip.

Sulfur cinquefoil flower *Source: Wise Co.*

Cinquefoil, Sulfur—*Potentilla recta* L.

Family: Rosaceae
Other common names: Old field five fingers, rough-fruited cinquefoil, upright cinquefoil
Life-cycle: Cool-season perennial
Native to: Eastern Mediterranean Eurasia

Distribution and Adaptation
- In the United States found in the eastern part of the country north into Minnesota, extending south to Georgia.
- Found in roadsides, vegetation disturbances, abandoned agricultural fields, and "waste areas"; cinquefoil can also invade native grasslands.
- Found in a variety of soil types (sand to clay soils) as well as on wet and dry lands.

Morphology/Growth Pattern
- Has an erect unbranched hairy stem growing from 1 to 2.6 ft (0.30−0.79 m) tall.
- Reproduces by seeds.

Use and Potential Problem
- Commonly a weed of pastures, turf, and landscapes.
- No feed value but sheep and deer are known to graze on sulfur cinquefoil; contains tannin which lowers its palatability to livestock.
- Due to its perennial nature and beautiful flowers, used in horticulture.

Toxicity
- None recorded.

Similar Species
- There are many Cinquefoils. Some have only three leaflets including *Rough Cinquefoil* (*Potentilla norvegica*) which has a hairy stem up to 39 in. (1 m) high and is a native. Rough cinquefoil has three coarsely toothed leaflets, while sulfur cinquefoil leaves are composed of five to seven leaflets and have smaller flowers than sulfur cinquefoil.
- *Dwarf mountain cinquefoil* (*Potentilla robbinsiana* L.) is often mistaken for strawberry. Plants with five leaflets include *Dwarf Cinquefoil* (*Potentilla canadensis*) which has basal leaves and leaflets toothed only beyond their middle and also mistaken for a strawberry. Also with five leaflets is *common cinquefoil* (*Potentilla simplex* L.), which is rather reclining with stems up to 2 ft (0.60 m) tall. Silvery Cinquefoil (*Potentilla argentea*) is nonnative and has small, silvery leaflets with only a few teeth and small flowers (less than 0.3 in. (1 cm) wide.

Reference
- Sulfur cinquefoil [19,32,33,39,47,66,67,72,74,76]

Cocklebur, Common—*Xanthium strumarium* L.

Common cocklebur seed bur

Brown spots on the stem

Common cocklebur seedhead/burs

Seed

- Size: 1.0−1.5 cm long, diameter 1.4−1.7 cm including spines and beaks
- Shape and texture: long, oval with many hooked spines and with two hooked teeth spreading from the apex; surface covered with many curly hairs, spiny, the bases of both the spines and the beaks are hairy
- Color: brown to black

Common cocklebur plant

Leaf shape and arrangement—Leaves: Alternate, long-petiolated, somewhat toothed and rough on lower and upper surfaces, sand papery leaves; 5−15 cm long and wide; stiff hairs on both sides of leaf surfaces, leaves with irregular margins.

Stolon/Rhizome/Roots: No stolons or rhizomes; has a rough green stem covered with distinct brown spots; stout taproot.

Inflorescence: Flowers small, male and female flowers borne in separate clusters; male flowers formed in clusters in the axils of the leaf; burs are oblong, covered with hooked spines and contain two fruits each with one seed.

Cocklebur, common—*Xanthium strumarium* L.

Family: Compositae
Other common names: Sheep bur, clot bur
Life cycle: Summer annual
Native to: Eurasia and Central America

Distribution and Adaptation
- Found throughout the United States. Particularly troublesome in southern states to Mexico. Most abundant in the Mississippi Valley.
- Most abundant in fertile soils, gardens, fields, roadsides, and other areas getting full sunlight.

Morphology/Growth Pattern
- Common cocklebur has branched, purplish green and rough-haired stems with distinctive prickly burs; the ridged stems often have purplish black spots; 1—5 ft (0.30—1.5 m) tall.
- When crushed, the plant gives off a distinctive odor.
- Reproduces by seed; the burs spread by hooking on clothing and the fur of animals.

Use and Potential Problems
- Primarily a weed of cultivated and reduced-tillage row crops. Also weed in pastures, waste places, roadsides, and stream banks.
- Two-leaf seedling stage poisonous to livestock; seeds and sometimes vegetation can be poisonous.
- Swine are the animals most affected (cause liver damage in pigs). Symptoms include vomiting and gastrointestinal irritation.

Toxicity
- Toxic, the main toxic principle is the glycoside carboxyatractyloside. It is concentrated in the seeds and seedlings. Mature plants are distasteful to animals and contain fewer toxins.
- Mechanical injury from prickly burs.

Similar Species
- *Cockleburs* are distinguished from each other based largely on variations in the shape, hairiness, and spines of the mature bur.

Reference
- Cockleburs [12,19,24,25,44,48,65,66,72,74,76]

Cudweed, Purple—*Gamochaeta purpurea* (L.) Cabrera

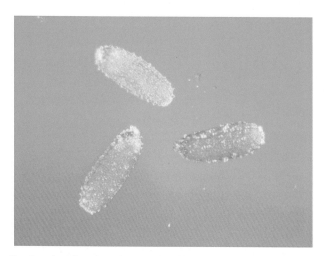

Purple cudweed seeds

Purple cudweed leaves *Source: NCSU*

Purple cudweed flowers *Source NCSU*

Seed
- Size: nutlets 0.5–0.7 mm long
- Shape and texture: elongated, flatter on one end; rough surface
- Color: yellowish to reddish brown

Purple cudweed plant *Source: NCSU*

Leaf shape and arrangement—Leaves: In a basal whorl, rosette; on the upper part of the plant, leaves alternate on the stem, tongue-shaped to oblong with blunt tips; 3–10 cm long, 1–2 cm wide; leaves covered with soft velvet-like hairs; upper leaf surface is dull green.

Stolon/Rhizome/Roots: No stolons or rhizome; stem highly branched from the base of the plant; stem covered with soft velvet-like hairs; taproot with a secondary fibrous root system.

Inflorescence: Flowers are tubular and are tinged brownish or purplish, flower heads clustered at upper leaf axils and terminal elongated cluster; bracts surrounding flower clusters are pink or purple in color; fruit produces bristly projections that shed at maturity.

Cudweed, purple—*Gnaphalium purpurea* L.

Family: Asteraceae
Other common names: Rabbit tobacco, purple everlasting, chafe-weed, catfoot
Life cycle: Summer or winter annual
Native to: Europe

Distribution and Adaptation
- European origin.
- Distributed throughout the United States but mostly found in the South.
- Prefers damp, acid soils.
- Productive under cool and moist conditions but cannot withstand hot and dry conditions.

Morphology/Growth Pattern
- Low-growing summer or winter annual; the young plant stems do not elongate and the leaves form a rosette; the stems of the mature plant, however, elongate from the rosette with or without branches; leaves on the elongated stems alternate, with grayish-green hairs; the flowers are small, spike-like, clustered in groups of 3−10 and located on the stem or at the base of leaf-stalk; flowers are tan to white and surrounded by light brown, pink, or purple bracts.
- Produces 100−500 seeds per plant.
- Reproduces by seed.

Use and Potential Problems
- No known feed values for livestock.
- Common weed of low maintenance turf, lawns, parks, and roadsides; less common in cultivated fields.

Toxicity
- None recorded.

Similar Species
- Similar species include *low cudweed* (*Galium uliginosum* L.) which is much shorter than purple cudweed and has much branching with smaller flowers.

Reference
- Purple cudweed [14,44,47,48,61,66,58,72]

Daisy, Ox-Eye—*Leucanthemum vulgare* Lam.

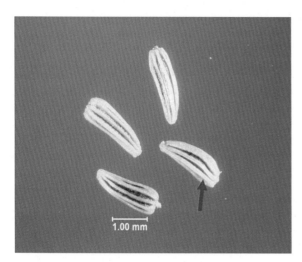

Ox-eye daisy seeds, white strips appear black

Ox-eye daisy clasping leaves

Oxeye daisy flower

Seed
- Size: 1.5–2 mm long, 0.6–0.9 mm wide
- Shape and texture: carrot- or comma-shaped, apex and base truncate (end abruptly), 8–10 tiny ridges; smaller than perennial sowthistle
- Color: white stripes, interspaces appear black

Oxeye daisy plant *Source John Wright*

Leaf shape and arrangement—Leaves: Lower leaves spoon-shaped, coarsely dissected and stalked; upper leaves narrower and stalkless (sessile/no petiole) or clasp the stem; leaf size progressively decreases upward on the stem; stem leaves 1.5–10 cm long, 2–15 cm wide.

Stolon/Rhizome/Roots: No stolons, has branched rhizomes and strong adventitious roots.

Inflorescence: Solitary flowers (daisy-like) at the end of branches, flowers composite with yellow centers, and 20–30 white petals radiating from the center, petals are slightly notched at the tip.

Daisy, ox-eye—*Leucanthemum vulgare* Lam.

Family: Asteraceae
Other common names: White daisy, field daisy, marguerite, poorland flower and moon-penny
Life cycle: Cool-season perennial
Native to: Europe

Distribution and Adaptation
- Found throughout North America, being most prevalent in the upper Mid-West.
- Ox-eye daisy prefers upland pastures and meadows, but also grows on roadsides, abandoned croplands, and waste areas.
- Found in waste ground, roadsides, and railroads.

Morphology/Growth Pattern
- Stems grow 1−3 ft (0.30−0.90 m) tall and are smooth, frequently grooved, and sometimes branch near the top.
- Ox-eye daisy looks like a typical daisy; flower heads are borne individually on the tops of long, slender stems.
- Form clumps or patches.
- Spreads by both seed and by short rhizomes.

Use and Potential Problems
- Is an aggressive invader of pastures, meadows, and roadsides throughout the United States. In western pastures and meadows, it can form dense stands which choke out other vegetation and decreases forage production.

Toxicity
- None recorded but may cause off flavor in milk if large amount consumed by cows.

Similar Species
- Often confused with the ornamental *Shasta daisy* (*Chrysanthemum maximum* L.) which is a more robust plant with larger flowers; central yellow disk flowers 2−3 cm wide; white ray flowers 2−3 cm long.

Reference
- Daisy, ox-eye [18,24,32,33,65,66,72,74,76]

Dandelion—*Taraxacum officinale* F.H. Wigg aggr.

Dandelion seeds. Rough projections on parachute

Dandelion flower *Source: Lachlan Cranswick*

Dandelion "parachute" of hairs *Source: James Altland*

Seed
- Size: 3.3—4.0 mm long, 1.1—1.3 mm wide
- Shape and texture: elongated, broad above middle and tapers to a round base and apex, apex has thin awn-like projection with parachute of hairs; surface dull, tiny ribs and spines on seed; rough projections on parachute end
- Color: greenish to reddish brown

Points on leaves point down; basal rosette

Leaf shape and arrangement—Leaves: Oblong in outline, rosette leaves, 5—40 cm long, leaves are deeply lobed, almost to the midrib, noticeably wavy with smaller lobes between the large ones, the terminal lobe is usually larger than the others; points on leaves point down.

Stolon/Rhizome/Roots: No stolons or rhizomes; erect, hollow flowering stems; deep taproot up to 1.3 cm in diameter.

Inflorescence: Single flower heads are on hollow, smooth stalks and have reflexed bracts; florets are all large, yellow in color and are solitary on the end of unbranched, leafless, hollow stalk.

Dandelion—*Taraxacum officinale* F.H. Wigg aggr.

Family: Compositae
Other common names: Lion-tooth, blow-ball, cankerwort
Life cycle: Cool-season perennial
Native to: Eurasia

Distribution and Adaptation
- Distributed throughout the United States.
- Grows in disturbed places, such as vacant lots, fallow fields, roadsides as well as undisturbed areas such as pastures, lawns, and meadows.

Morphology/Growth Pattern
- Dandelion has a rosette growth habit; leaves are deeply lobed almost to mid-rib, with the smaller lobes between the larger ones. The single heads are on hollow, smooth stalks and have reflexed bracts; plant with a single flower head can grow up to 2.5 ft (0.76 m) tall.
- Reproduces by seed and by new growth from the root crown; new growth can also arise from the bud on the fleshy taproot following damage to the plant (such as cultivation); the taproot of chicory responds in a similar manner.

Use and Potential Problems
- Dandelion is one of the most common weeds of turfgrass and lawns throughout the United States. Dandelion also occurs as a weed of container ornamentals, landscapes, nurseries, orchards, and occasionally agronomic crops.
- Used for rabbit fodder and also used in cat and dog foods.
- Has a high forage value when vegetative.

Toxicity
- None recorded.

Similar Species
- At the rosette stage dandelion and *chicory* (*Cichorium intybus* L.) resemble each other; dandelion leaves are arranged opposite from each other and point toward the rosette, while those of chicory are not always arranged opposite and point either toward the center of the rosette or away from the center of the rosette; chicory also has blue flowers and a flowering stem; in addition to chicory, two other species also resemble dandelion: *field hawkweed* (*Hieracium caespitosum* L.) and *Cat's-Ears* (*Hypochoeris radicata* L.); field hawkweed unlike dandelion does have toothed margins, also field hawkweed has flower stalks that are covered with stiff and dense hairs; false dandelion has yellow flowers on tall, leafless stems and its leaf is unlike dandelion; irregular to rounded lobes and is hairy.

Reference
- Dandelion [1,18,32,33,44,47,66,67,72,74,76]

Dayflower, Climbing—*Commelina diffusa* Burm. f.; Dayflower, Common or Asiatic—*Commelina diffusa* L.

Dayflower seeds

Seed
- Size: 2–4.5 mm long
- Shape and texture: seeds have small depressions or pits
- Color: light gray

Dayflower plant *Source: Nathan O'Berry*

Dayflower flower

Leaf shape and arrangement—Leaves: Alternate on the stem, ovate, narrow, long, and pointed with parallel veins; 3–15 cm long, 1–3.5 cm wide; the leaf wraps around the main branches and forms sheaths at the base.

Stolon/Rhizome/Roots: No stolons or rhizomes; stems thick, reaching 6.35 cm in length, swollen at the nodes, and often rooting when nodes come into contact with soil; has a fibrous root systems.

Inflorescence: Flowers appear in cluster, each flower has three sepals and three petals, two of the petals are blue and rounded and much larger, while the third one is smaller and white; flower also contains six yellow unequal size stamens; the bracts are green and heart-shaped.

Dayflower, climbing *Commelina diffusa* Burm. f.

Dayflower, common or Asiatic—*Commelina diffusa* L.

Family: Commelinaceae
Other common names: Common dayflower, wandering jew
Life cycle: Summer annual
Native to: Asia

Distribution and Adaptation
- Grows from the New England states south to Florida and westward to Kansas and Texas.
- Dayflower is commonly found in moist shady places.

Morphology/Growth Pattern
- An erect or more often creeping annual monocot often mistaken for a broadleaf weed due to its attractive blue flowers. Can grow up to 1–3 ft (0.30–0.90 m) tall.
- Foliage resembles grass leaves due to their parallel veins and lack of petioles.
- Annual has fleshy, succulent branches that creep and spread by rooting at the nodes.
- The plant gets its name from the unusual fact that each flower lasts only 1 day.
- Reproduces by seed.

Use and Potential Problems
- Mostly considered a weed.
- Found at roadsides, railroads, gardens, barnyards, fields, fallow and cultivated fields.

Toxicity
- None recorded.

Reference
- Common or Asiatic dayflower [24,46,58,66,72]

Deadnettle, Purple or Red—*Lamium purpureum* L.

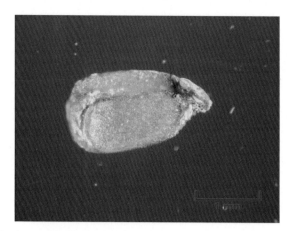

Purple or red deadnettle seed

Left to right henbit and purple deadnettle *Source: NCSU*

Purple deadnettle flower head *Source: NCSU*

Seed
- Size: 2.5−3 mm long and 1−1.5 mm wide
- Shape and texture: oval shaped
- Color: light brown mottled with whit spots

Purple deadnettle plant *Source: NCSU*

Leaf shape and arrangement—Leaves: Arranged opposite from each other; lower leaves with petioles; petioles 3 cm long, upper leaves are sessile and becomes triangular and pointed; sparsely pubescent, deep-green to purplish; leaf 0.5−3.5 cm long and broad; leaves generally are less than twice as long as broad.

Stolon/Rhizome/Roots: No stolons or rhizomes; stem decumbent at base to erect, hollow and four-angled (square); stem branched from the base and sometimes from the axils; up to 45 cm tall; fibrous root system.

Inflorescence: Flowers are tubular and sessile, the tubular part of the flower extends beyond calyx tip; the pinkish-purple flower's lowest petal is divided into two diverging lobes and has dark colored spot, the flowers appear in whorls at base of upper leaves.

Deadnettle, Purple or Red—*Lamium purpureum* L.

Family: Lamiaceae
Other common names: dead nettle, blind nettle, Morningglory, pitted: white flowered morningglory, small morning-glory https://extension.tennessee.edu/publications/Documents/W124.pdf and Mullein, moth: non found
Life cycle: Winter annual
Native to: Europe

Distribution and Adaptation
- Found throughout the United States but not as abundant as *henbit* (*Lamium amplexicaule* L.).
- Found in waste ground, roadsides, railroads, pastures, and low maintenance turf.

Morphology/Growth Pattern
- This low growing plant has greenish to purplish, tender, square stems, the stem branches from the base; leaves are heart-shaped, crowded, overlapping, and arranged in whorls, leaves have distinct deep veins, lower leaves are on long petiole, while upper leaves are on short petioles (compared with henbit with no petioles on upper leaves); leaves often purple-tinged; flowers are reddish pink to purple in color and are tubular; can reach up to 13−1.5 ft (0.40−0.46 m) in height.
- Spreads by seed and by rooting at internodes.

Use and Potential Problems
- Purple deadnettle has no known feed values for livestock; is mostly a weed of pastures, turf, gardens, and landscape.

Toxicity
- None recorded.

Similar Species
- Purple deadnettle is often confused with henbit (*L. amplexicaule* L.). Unlike purple deadnettle, henbit leaves are more pubescent and their upper leaves do not have petioles (sessile), the leaves clasp the stems; the leaves of purple deadnettle are more triangular and less deeply lobed than henbit; purple deadnettle flowers are lighter purple than henbit flowers.

Reference
- Purple or red deadnettle [14,32,33,65,66,72,74,76]

Dock, Curly—*Rumex crispus* L.

Curly dock seeds

Curly dock leaf *Source: NCSU*

Curly dock seed heads *Source: NCSU*

Seed
- Size: 2.1—2.6 mm long, 1.4—1.7 mm wide; larger than red sorrel, may be *shown in hull*
- Shape and texture: ovate, three-angled and the angles are narrowly winged with sharp ridges; surface smooth, very shiny
- Color: reddish brown

Curly dock plant

Leaf shape and arrangement—Leaves: Alternate, simple, lanceolate, prominently curly and wavy along the margins and mostly basal; 10—30 cm long; stipules fused to form a sheath above each node (ocrea); upper leaves' petioles short or absent.

Stolon/Rhizome/Roots: No stolons or rhizomes; yellowish orange in color; enlarged node and each node is surrounded by a membranous sheath; large, fleshy, branched taproot.

Inflorescence: Flowers in several whorls crowded into a cluster of ascending racemes; numerous winged (whorls) seedpods, green (young), brown (mature); bisexual with six sepals, the inter three developing wings of the fruit (forming three-winged triangular structure each with a swelling or growth); no petals, papery seedhead; spearlike leaves in seedhead.

Dock, curly—*Rumex crispus* L.

Family: Polygonaceae
Other common names: Sour dock, yellow dock, narrow-leaved dock
Life cycle: Cool-season perennial
Native to: Europe

Distribution and Adaptation
- Distributed throughout the United States and southern Canada. Also in West Indies, Mexico, and Central America, Asia, and Australia.
- Found in meadows, pastures, lawns, and waste places.
- Fields, roadsides, railroads, waste ground, disturbed sites, and wet habitats.

Morphology/Growth Pattern
- Curly dock is a robust perennial plant, having a reddish, unbranched stem; during the first year, the plant forms a dense rosette of leaves on a stout, fleshy taproot with yellow center; plant can grow up to 5 ft (1.6 m).
- Reproduce by seeds.

Use and Potential Problems
- Curly dock is a weed of low maintenance turfgrass, pasture, some row *crops*, roadsides, railroads, waste ground, disturbed sites, and wet habitats.
- Plants can contain high levels of oxalic acid, which tie-up nutrients such as calcium in food causing mineral deficiencies. Oxalic acid gives the plant an acid-lemon flavor.
- Important source of food for the caterpillars of many butterflies.

Toxicity
- Oxalates.

Similar Species
- Curly dock is often confused with *broadleaf dock* or *bitter dock* (*Rumex obtusifolius* L.), however, broadleaf dock or bitter dock has leaves that are much wider and lack curly margins and the leaves have heart-shaped lobes at the base.

Reference
- Curly dock [1,12,14,18,25,32,33,44,48,65,66,72,74,76]

Dodder—*Cuscuta* L. spp.

Dodder seeds

Dodder on alfalfa plant *Source: Nathan O'Berry*

Seed
- Size: 1.1–1.6 mm long, 0.9–1.3 mm wide, varies from size of alsike clover to red clover
- Shape and texture: nearly round with one side round and two flat sides, hilum is in the truncate or notched area at the base of the ventral side; dull, rough surface
- Color: yellow to light brown

Dodder plant *Source: Nathan O'Berry*

Leaf shape and arrangements—Dodder is leafless, slender, with string-like stems that twine all around the host plant.

Stolon/Rhizome/Roots: No stolons or rhizomes; thread-like, leafless stems twine around host plants; stems may be yellow, orange, pink, or brown in color and are only 0.3–1.5 mm thick.

Inflorescence: Flowers are small, white or yellow, bell-like; can be borne in tight balls or in a loose cluster; as the flower matures, they produce tiny seeds. Seeds in a round capsule about 3 mm long, which contains four seeds.

Dodder—*Cuscuta* L. spp.

Family: Convolvulaceae/Cuscutaceae
Other common names: Strangleweed, devil's hair, love vine, hell-bind, gold thread, and yellow stringy plants, field dodder
Life cycle: Cool-season annual
Native to: Europe

Distribution and Adaptation
- Many different species of dodder found throughout the United States.
- Tolerates wet sites, the seed often gets into irrigation water and is carried to the fields.
- Different species of dodder are host specific: *Cuscuta salina* L. is found in salty marshes, flats, and ponds on just a few host plants species; *Cuscuta pentagona* on crop and weed species including alfalfa (*Medicago sativa* L.), asparagus (*Asparagus officinalis* L.), safflower (*Carthamus tinctorius* L.), sugarbeet (*Beta vulgaris* L.), pigweed (*Amaranthus* spp.), lamb's-quarters (*Chenopodium album* L.), and field bindweed (*Convolvulus arvensis* L.). *Cuscuta indecora* hosts include alfalfa and many weeds such as Russian thistle (*Salsola tragus* L.), field bindweed, lamb's-quarters, and five-hook bassia (*Bassia hyssopifolia* L.).

Morphology/Growth Pattern
- Dodder is a parasite that exists on host plants. Dodder does not produce roots, leaves, or green coloring matter (chlorophyll). The seed germinates in or on the soil, and sends out a small thread-like stem; if it does not find a favorable host plant to attach itself within a week, it collapses and dies. Once the dodder is attached to the host plant the root shrivels and dies. The slender stem grows in a spiral manner around the host plant and produces sucking organs (haustoria) that take water and minerals from the host plant. As the dodder continues to grow, new haustoria are formed and they settle firmly into the host.
- As the branching network of the dodder expands, it attacks other nearby plants. The dodder causes the host plant to be weakened and discolored after living off of it for some time. Dodder seedlings must attach to a suitable host within a few days of germinating or they die.
- The vining nature and yellowish to reddish-brown color of dodder makes it easy to distinguish from most other weeds, however, several species of dodder occur throughout the United States.
- Reproduces by seed, can also reproduce vegetatively from detached stem pieces.
- Produces tiny seeds that are viable (able to germinate) for up to 5 years.

Use and Potential Problems
- Dodder is a parasitic plant that causes considerable damage to forages and some field crops; especially destructive to alfalfa, lespedeza, flax, clover, and potatoes.
- Dodder is particularly troublesome where alfalfa, clover, and onion are grown for seed because dodder seed is difficult to remove from the desired seed crop and can be spread with infested seed.
- Considered a weed of landscape, nursery crops, and agricultural crops. Dodder only survives if it attaches itself to an appropriate host plant.

Toxicity
- Various toxins.

Similar Species
- *Pine mistletoe* (*Arceuthobium americanum* Nutt.) is similar to dodder, however, pine mistletoe is found mainly on branches of pines; the yellowish green stem has numerous branches and small green flowers.

Reference
- Dodder [12,18,25,48,65,66,72]

Filaree, Redstem—*Erodium cicutarium* (L.) L'Hér <u>ex</u> Ait.

Redstem filaree plant with fruits

Redstem filaree plant *Source: James Altland*

Redstem filaree flowers

Leaf shape and arrangement—Leaves: Pinnately compound and deeply dissected (lobed) into numerous segments; leaves opposite on upper portion of the stem and alternate below; basal leaves are petiolated, stem leaves are sessile, individual leaves are divided into three to nine leaflets, 5–20 cm in size; hairy, reddish in color, fernlike, or feathery appearance; plants form a rosette close to the ground.

Stolons/Rhizomes/Roots: No stolons or rhizomes; stems are reddish; shallow taproot with secondary fibrous root systems.

Inflorescence: Flowers arranged on a long stalk (pedicel); clusters of two to eight flowers occur with each individual flower; individual flowers 1.3 cm wide consisting of five bright pink to purple petals; fruit 1.3–2 cm long, sharp-pointed, narrow capsule resembles a bird's beak.

Filaree, redstem—*Erodium cicutarium* (L.) L'Hér <u>ex</u> Ait.

Other common names: Stork's bill, pin clover, alfileria, filaria, heron's bill, pin-weed, pin-grass, cutleaf filaree, purple filaree cranesbill
Family: Geraniaceae
Life cycle: Winter annual or biennial
Native to: Mediterranean region of Europe or Asia

Distribution and Adaptation
- Found in most of the United States, excluding Florida.
- Occupies a variety of habitats, from desert to riparian.
- Grows in well-drained, clayey, loamy, or sandy soil.
- Tolerates broad range of climates.

Morphology/Growth Pattern
- Prostrate to semierect winter annual or biennial with numerous branches rising from the basal rosette, a member of the geranium family.
- Older leaves grow up to 1 foot (0.30 m) long.
- Often referred to as stork's bill because the seed awns look like the beak of a stork.
- The awns coil while drying and open when dampened, which allows the seed to bore into the soil.
- Reproduces by seed.

Use and Potential Problems
- Redstem filaree is primarily a weed of turfgrass and landscapes, also found in small grain and alfalfa fields.
- Known to provide seasonal forage for rodents, desert tortoise, big game animals, and livestock; has been reported to cause bloat.

Toxicity
- None recorded.

Similar Species
- The seedheads of *Carolina geranium* (*Geranium carolinianum* L.) resemble redstem filaree; however, the leaves of Carolina geranium are rounded and palmately veined; also the fruit beaks of Carolina geranium are coiled outwardly at maturity while the redstem filaree beaks are spirally twisted when dry, with a few spreading trichomes.

Reference
- Redstem filaree [14,19,44,58,68,72,74]

Fleabane, Annual—*Erigeron annuus* (L.) Pers.

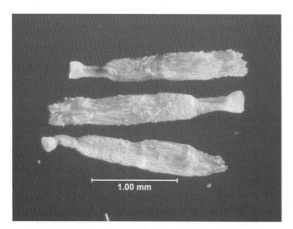

Annual fleabane seeds

Daisy fleabane plant *Source: Lachlan Cranswick*

Annual fleabane stem and leaf *Source: Rocky Lemus*

Leaf shape and arrangement—Leaves: Egg-shaped (ovate) or widest near the middle and tapering to both ends (lanceolate), seedling leaves appear in basal rosette, coarsely toothed, stalked (*petioled*), bright green and usually sparsely hairy; mature leaves narrow, alternate on the stem, upper leaves short petioled or sessile clasping the stem at the broadened base, shallowly toothed and much smaller than the basal leaves.

Stolons/Rhizomes/Roots: No stolons or rhizomes, stem erect, solid, branched with many soft hairs; root varies from taproot to fibrous.

Inflorescence: Flowers arranged in a cluster at the end of stems and branches forming a flat-topped inflorescence, white or purplish outer ray-florets with inner core yellow.

Daisy fleabane flower *Source: Lachlan Cranswick*

Fleabane, annual—*Erigeron annuus* (L.) Pers

Other common names: Daisy
Family: Asteraceae
Life cycle: Summer annual or sometimes biennial
Native to: Europe

Distribution and Adaptation
- Found across much of the United States except the extreme north, also found in southern Canada.
- Prefers full or partial sun, and moist to slightly dry conditions.
- Grows on variety of soils, and tolerates clay or gravel more so than similar weeds.

Morphology/Growth Pattern
- The plant grows up to 3.5 ft (1 m) tall, branching occasionally to form flowering stems.
- Reproduces by seed.

Use and Potential Problems
- Primarily a weed of abandoned fields, pastures, roadsides, railroads, disturbed open woods, and various kinds of waste areas.

Toxicity
- None recorded.

Similar Species
- Annual fleabane is often confused with *daisy fleabane* (*Erigeron strigosus* L.), but daisy fleabane differs from annual fleabane by its more numerous and broader leaves (especially toward the upper part of stems), and the long, spreading white hairs that occur along the stems compared to the much shorter, appressed hairs found along the stems of daisy fleabane.

Reference
- Annual fleabane [44,47,58,72]

Galinsoga, Hairy—*Galinsoga quadriradiata* Ruiz & Pavon, Previously Known as *Galinsoga ciliata* (Raf.) Blake)

Hairy galinsoga seeds

Hairy galinsoga leaves and flowers *Source: Lachlan Cranswick*

Hairy galinsoga flower *Source: Lachlan Cranswick*

Hairy galinsoga plant *Source: Lachlan Cranswick*

Leaf shape and arrangements—Leaves: Lance-shaped (with a pointed tip); opposite, whole and coarsely toothed; the upper surface of the leaves, stems, and petioles are densely covered with hairs, on the lower surfaces of the leaf, however, hairs only appear on the leaf veins, leaf blades with three prominent veins; the leaves are long-stalked (petioled) on the lower part of the stem but become almost sessile on the upper flowering branches.

Stolons/Rhizomes/Roots: No stolons or rhizomes; hairy erect stems, freely branched, the stems are sometimes maroon-tinted; fibrous root systems.

Inflorescence: Flower composed of five small white rays, each one of the white rays are three-toothed, arranged around a yellow central head, 0.5 cm wide.

Galinsoga, hairy—*Galinsoga quadriradiata* Ruiz & Pavon, previously known as *Galinsoga ciliata* (Raf.) Blake)

Other common names: Peruvian daisy, galinsoga quadriradiata, common quickweed, shaggy soldier
Family: Compositae/Asteraceae
Life cycle: Summer annual
Native to: Central and South America.

Distribution and Adaptation
- Found throughout the eastern and Midwestern United States and also on the west coast.
- It is an annual plant, requires recently exposed soil for good establishment, so while it is found in gardens and crop fields, along roadsides, and other disturbed places, it is not particularly invasive of natural plant communities.

Morphology/Growth Pattern
- Freely branched, spreading, erect, and pubescent.
- Grows between 6 and 18 in. (15−45 cm) tall.
- Reproduces by seed.

Use and Potential Problems
- Primarily a weed of vegetable crops, however, it may occur in any cultivated situation.
- Under moist conditions can become a serious problem under irrigation or in areas that receive year-round rainfall.

Toxicity
- None recorded.

Similar Species
- *Smallflower galinsoga* (*Galinsoga parviflora* Cav.,) is similar *to hairy galinsoga*, hairy in appearance, however, the stems of smallflower galinsoga are smooth to sparsely hairy unlike hairy galinsoga which is densely hairy; another distinguishing characteristic between these two species is the pappus (calyx) of the disk-florets. In hairy galinsoga, the calyx of the tubular flowers tapers to a very fine tip (awn) while in smallflower galinsoga the calyx (pappus) is awnless, also smallflower galinsoga lacks the pappus.

Reference
- Hairy galinsoga [47,58,72]

Garlic, Wild—*Allium vineale* L.

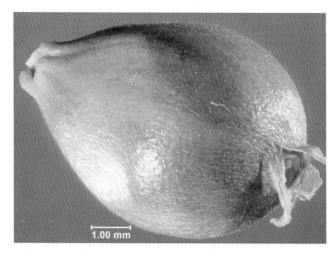

Wild garlic seed

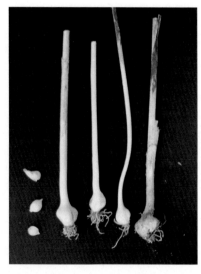

Stem, leaves, and underground bulb *Source: NCSU*

Wild garlic aerial and underground bulbs

Seed/Bulb
- Like tiny garlic or onion bulblets
- Bulb located on base of stalk, long peduncle
- Paper-like bulblets enclose a black seed

Wild garlic plant *Source: NCSU*

Leaf shape and arrangement—Leaves: Basal leaves are few and originated from the base of the stem (bulb), leaves are linear, slender, long, pointed, hollow, nearly round, and are attached to the lower half of the stem; 15−60 cm long and 0.2−1 cm wide.

Stolon/Rhizome/Roots: No stolons or rhizomes; flowering stems are slender, smooth, waxy, solid, unbranched, erect, and leafless.

Inflorescence: Aerial bullets form in oval, smooth, shiny clusters at the top of the stem, and produces long, green, thread-like leaves; flowers are red, purple, or pink; flowers may be replaced by aerial bulblets.

Garlic, wild—*Allium vineale* L.

Other common names: Field garlic, scallions, wild onion, crow garlic
Family: Liliaceae
Life cycle: Cool-season, perennial from bulbs
Native to: Europe

Distribution and Adaptation
- Found throughout most of the eastern and southern United States and also found in Canada.
- Drought tolerant and can grow on wide range of soil conditions.

Morphology/Growth Pattern
- The plant grows from underground, onion-like but becomes flattened; leaves occur on the flowering stem up to half the height of the plant; stem slender and hollow up to 3 ft (0.90 m) tall bearing a terminal cluster of flowers in umbel and may produce many greenish or purplish flowers. Underground white bulb bears bulblets that are flattened on one side and enclosed by thin papery-like membrane. Bulbs will shed to form new plants or sometime will sprout in the head to form a bushy mass of green seedlings.
- Has both underground and aerial bulblets.
- A strong odor is associated with all parts of the plant.
- Reproduces by aerial and underground bulb offsets, aerial bulblets, and seed.

Use and Potential Problem
- Wild garlic was introduced by early settlers as a food flavoring; it is a common weed of pastures, turfgrass, nursery crops, landscapes, winter wheat, and other cultivated crops.
- If seed harvested with wheat, bread made of the flour can taste garlicky.
- Wild garlic like wild onion, when eaten by livestock or poultry, taints meat, milk, and poultry products.

Toxicity
- Sulfur containing compounds.

Similar Species
- Wild garlic is often confused with *wild onion* (*Allium canadense* L.). Both plants often occupy the same sites. Wild onion can be distinguished from wild garlic by its leaves that are flat in cross-section and not hollow; the presence of a fibrous coat on the central bulb, the offset bulblets at the base of the plant, and the areal bulblets readily distinguish wild garlic from wild onion. Also, in wild onion, the bulbs do not divide to form a new plant as in the case of wild garlic.

Reference
- Wild garlic [12,14,24,33,44,47,67,72,74,76]

Geranium, Carolina—*Geranium carolinianum* L.

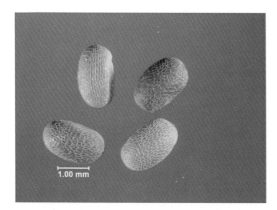

Carolina geranium seeds

Carolina geranium leaf *Source: NCSU*

Carolina geranium flowers *Source: NCSU*

Seed
- Size: 1.7–2 mm long, 1.1–1.3 mm wide
- Shape and texture: ovate or rounded oblong with a small tip on one end; rough-textured surface, network of fine ridges from hilum to the opposite end of seed, light veins
- Color: dark red-brown or gray-brown

Carolina geranium plant *Source: NCSU*

Leaf shape and arrangement—Leaves: Round in appearance, and are deeply divided into five to nine segments and each segment is further lobed or toothed; alternate near base, opposite above, leaf on long *petioles*; hairy on both leaf and stem surfaces.

Stolons/Rhizomes/Roots: No stolons or rhizomes; erect, freely branching near base, up to 71 cm tall; the stems are usually pink to red in color and densely hairy; fibrous root system.

Inflorescence: Flowers appear in two or more clusters on the tips of the stem; petals are whitish-pink; fruit is a five-parted capsule that forms a "stork's bill" up to 1.3 cm long.

Geranium, Carolina—*Geranium carolinianum* L.

Other common names: Cranesbill, wild geranium, Carolina crane's-bill
Family: Geraniaceae
Life cycle: Winter annual/biennial
Native to: Europe

Distribution and Adaptation
- Although it thrives in warmer areas, it grows throughout the United States and southern Canada.
- Grows in thinner areas of the lawn and pastures.

Morphology/Growth Pattern
- Low-growing and spreading, up to 1.8 ft (0.55 m) tall.
- Most often a biennial, forming a basal rosette initially with subsequent stem elongation and branching as the plant matures.
- Flowers are red-violet, and seeds are smooth. The ovaries of Carolina geranium and small flower geranium are hairy; the ovaries of dovefoot geranium are wrinkled and lack hairs.
- Reproduces by seed.

Use and Potential Problems
- Mostly turfgrass and pasture weed.
- Occurs in disturbed areas, gardens, cultivated fields, pastures, roadsides, and waste places.
- The seeds are eaten by birds and rodents.

Toxicity
- None recorded.

Similar Species
- Its deeply divided leaves and distinct fruit shape easily distinguishes Carolina geranium from other geranium species such as *Dovefoot geranium* (*Geranium molle* L.), also known as cranesbill, and *smallflower geranium* (*Geranium pusillum* L.) also known as smallflower cranesbill, which are biennials with rounded leaves not as deeply divided as those of Carolina geranium.

Reference
- Carolina geranium [19,24,33,47,61,67,72,76]

Goldenrod, Canada or Common—*Solidago canadensis* L.

Goldenrod seed

Goldenrod leaves

Goldenrod flower

Seed

- Size: 1—1.5 mm long
- Shape and texture: oblong or five-sided; surface with closely spaced ridges; pubescent with pappus 3—4 mm long
- Color: ridges are pale tan and the interspaces brown

Goldenrod plant *Source: Edward McCann*

Leaf shape and arrangement—Leaves: Narrowly lanceolate in shape, tapering to the base and tip; alternate on the stem, are sessile, up to 3—15 cm long and 0.5—2.2 cm wide with three-linear veins; leaves with toothed margins; finely hairy on the underside, rough to nearly smooth on the upper surface.

Stolons/Rhizomes/Roots: No stolons but has long creeping rhizomes; stem hairy in the upper portion of the plant, and smooth below; the rhizomes primarily arise from the aerial stem; the reddish rhizomes can reach 5—12 cm long; plant with extensive fibrous root system.

Inflorescence: The yellow flowers which head on arching branches are located in a long flat-tapped terminal panicle; the flowers are tightly packed, and each flower has 9—15 rays and approximately 7 central flowers with 5 petals; the bracts underneath the flowers are linear and yellow with green pointed tips.

Goldenrod, Canada or common—*Solidago canadensis* L

Other common names: Common goldenrod
Family: Compositae/Asteraceae
Life cycle: Warm-season perennial
Native to: United States and Canada

Distribution and Adaptation
- Goldenrod grows under a wide range of soil and climatic condition.
- Often found in fertile soils as well as periodically waterlogged but not permanently wet or extremely dry areas.

Morphology/Growth Pattern
- A tall, erect, rhizomatous perennial that grows over 1−3.3 ft (0.30−1 m) tall.
- Grows in patches, stems arise from the base of the plant in singles or loosely clustered; stems short, pubescence in the upper part of the plant; leaves numerous and are distributed throughout the stem; leaves are attached to the stem (sessile), narrow, 1.2−6 in. (3−15 cm) long, three-nerved, entire to serrate, smooth to slightly pubescent; yellow flowers numerous and heads in a terminal panicle. By the time the plant flowers, the lower leaves are usually gone.
- Reproduces by seed and creeping rhizomes.

Use and Potential Problems
- Classified as a weed; seldom found in cultivated fields; common in pastures, nursery crops, orchards, roadsides, and meadows.
- Can be controlled by rotational grazing.
- Not preferred by many livestock, except goats and sheep.
- Plant provides protection for wildlife; seed is consumed by some birds and white-tailed deer.

Toxic
- None recorded.

Similar Species
- Canada goldenrod now includes *tall goldenrod* (*Solidago canadensis* var. *scabra*), tall goldenrod was previously considered as a separate species (*Solidago altissima* L.). The two species resemble each other; however, tall goldenrod has grayish leaves and uniform hairs on the stem, while Canada goldenrod has green leaves and no hairs on the bottom third of the plant. Another species, *Missouri goldenrod* (*Solidago missouriensis* L.), is also similar to Canada goldenrod, but Missouri goldenrod is usually smaller with smooth stems.

Reference
- Canada or common goldenrod [34,39,46,47,58,68,72]

Gromwell, Corn or Field—*Buglossoides arvensis* (L.) I.M. Johnston

Field/corn gromwell seeds

Fruit: 4 small conical nutlets at leaf axis

Corn gromwell flowers

Seed

- Size: 2.6–3.7 mm long, 1.7–2.6 mm wide
- Shape and texture: shaped like pear, with the narrow end bent to one side and the wider end is truncate; distinct ridge from the base to the pointed apex, small projections aligned with the ridge; surface covered with small tubercles and wrinkles
- Color: light gray-brown to brown overall, wrinkles and tubercles are light

Corn gromwell plant

Leaf shape and arrangement—Leaves: Lanceolate to linear, 1–3 cm long, sessile, rough, with one vein, pale green, covered with small persistent hairs, attenuated at base and toward top.

Stolons/Rhizomes/Roots: No stolons or rhizomes; stem erect or ascending, slightly branched, stem covered with dense hairs; has well-developed root system.

Inflorescence: White small flowers in slightly curved raceme, located in the axils of the leaves; white flowers, the petals the same length as the sepals; fruits brownish, roughly covered by small tubers or projections.

Gromwell, corn or field—*Buglossoides arvensis* (L.) I.M. Johnston

Other common names: Bastard alkanet
Family: Boraginaceae
Life cycle: Winter annual or biennial
Native to: Europe

Distribution and Adaptation
- In the United States, distributed throughout most of the states.
- Corn gromwell can be found along roadsides, in fields, and other disturbed places.

Morphology/Growth Pattern
- Leaves have entire margins, no petioles and have flat, stiff, short hairs on both surfaces of the leaf; stem erect, slender, single stem or several stems branching from the base; if branching, the central stem is larger than the rest; flowers are creamy-white and occur in the reduced leaf axils; individual flowers are 5–8 mm long, funnel-shaped.
- Plant height varies greatly with the habitat—4–28 in. (10–70 cm).
- Reproduces by seed.

Use and Potential Problems
- Livestock feed values not known.
- The plant grows along roadsides and in old fields.
- Troublesome weed in small grains.

Toxicity
- None recorded.

Reference
- Corn or field gromwell [19,47,66,74]

Groundsel, Common—*Senecio vulgaris* L.

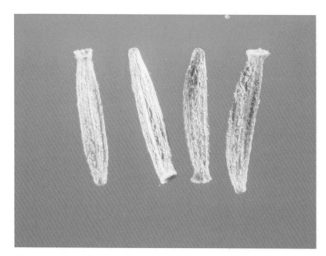

Common groundsel seed

Common flower heads *Source: James Altland*

Yellow flower

Seed
- Size: 2.2−2.5 mm long, 0.4−0.5 mm wide
- Shape and texture: seeds are long and narrow, pappus of fine white hair 5−6 mm long is attached to the upper end of the seed
- Color: tan

Common groundsel plant *Source: James Altland*

Leaf shape and arrangement—Leaves: Oblong or lanceolate and are deeply, coarsely, irregularly toothed/lobed or pinnately parted; also has blunt teeth along the margins of the leaf; the edge of the leaf margins are usually curved downward; alternate, the lower leaves have short, stout *petioles* while the upper leaves are sessile and are clasping; hairs on leaves variable ranging from nearly smooth to densely covered with white hairs that are cobwebb-like in appearance; the stem and basal leaves usually purplish underneath; leaves 2.5−10 cm long.

Stolons/Rhizomes/Roots: No stolons or rhizomes; the hairs on the stem and leaves are variable, ranging from nearly none to dense white hairs; round, hollow stems often have fine veins or longitudinal ridges; small taproot and a fibrous root system.

Inflorescence: Upper flowering stems terminate in small, dense clusters of composite flowers, each composite flower is more or less cylindrical in shape, spanning about 0.5−1 cm in diameter and 1 cm in length; each disk floret has five slender lobes.

Groundsel, common—*Senecio vulgaris* L.

Other common names: Grimsel, simson, bird-seed, ragwort
Family: Compositae/Asteraceae
Life cycle: Winter annual
Native to: Europe and Asian

Distribution and Adaptation
- Distributed throughout the United States, but most common in the northern part of the United States, Texas, and California; also found in Canada.
- Commonly found in pastures, field nursery crops, greenhouses, landscapes, less often in cultivated field crops.
- The genus *Senecio* is one of the largest genera of plants, containing over 1000 species.

Morphology/Growth Pattern
- An upright or reclining, somewhat succulent, usually branching, glabrous plant; leaves alternate on the stem, deeply and irregularly lobed, smooth and slightly fleshy; lower leaves on short petiole and upper leaves sessile, partly clasping the stem; flowers clustered, heads cylindrical, with smooth bracts and yellow, tubular flowers as long or slightly longer than the surrounding bracts; small taproot with secondary fibrous root system.
- Common groundsel can grow up to 2 ft (0.60 m) tall.
- Reproduces by seed; is an extremely prolific seeder, continually producing large numbers of seeds.

Use and Potential Problems
- No livestock feed values; troublesome weed in moist areas.
- Contain alkaloids that make it poisonous to horses and cattle; sheep have a much higher tolerance.

Toxicity
- Pyrrolizidine alkaloids.

Similar Species
- Common groundsel leaves resemble seedlings of *common ragweed* (*Ambrosia artemisiifolia* L.), but common ragweed leaves are much more deeply seriated than groundsel.

Reference
- Common groundsel [6,12,47,58,72,74]

Hawksbeard, Smooth or Redstem—*Crepis capillaris* (L.) Wallr.

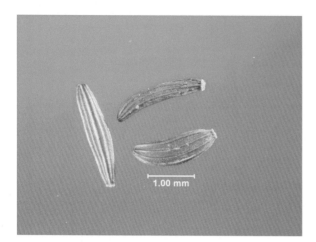

Redstem hawksbeard seeds

Seed
- Size: 2.2—2.3 mm long, 0.5—0.7 mm wide
- Shape and texture: 10 ribs on surface, glabrous, pappus 3—4 mm long
- Color: light to dark brown

Red stem hawksbeard plant

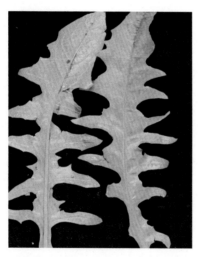

Red stem hawksbeard leaves

Leaf shape and arrangement—Leaves: Lanceolate in outline, approximately 6—20 cm long, 1.3—7.6 cm wide, leaves are usually dissected or lobed, the upper leaves clasp the stem with a pair of backward-pointing basal lobes, while the basal leaves occur on petioles.

Stolons/Rhizomes/Roots: No stolons or rhizomes; stems are erect, reaching 8.5 cm in height, without hairs, branching from the base.

Inflorescence: Flowers occur at the end of the branches; individual flowers are 0.8—1.0 cm long, the bracts around the flower heads have longer inner rows and much shorter outer rows.

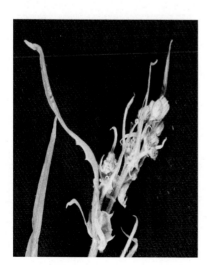

Red stem hawksbeard seedhead

Hawksbeard, smooth or redstem—*Crepis capillaris* (L.) Wallr.

Other common names: None found
Family: Compositae/Asteraceae
Life cycle: Cool-season *biennial or perennial*
Native to: Europe

Distribution and Adaptation
- In the United States found throughout the south.

Morphology/Growth Pattern
- Erect plant, has lanceolate leaf shape with deeply lobed leaves that clasp at the base; the base leaves have petioles, while the leaves on the upper part of the stem are sessile (directly attached to the stem).
- Reproduces by seed.

Use and Potential Problems
- Primarily a weed of pastures, hay fields, and roadsides.
- No known livestock feed values.
- Seedheads and leaves often consumed by white-tailed deer.

Toxicity
- None recorded.

Reference
- Smooth or redstem hawksbeard [33,47,58,61]

Hawkweed, Field—*Hieracium* L. spp.

Field hawkweed seeds

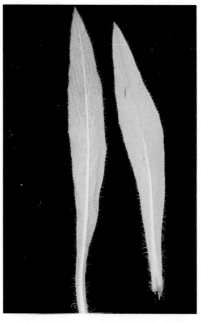

Yellow hawkweed leaves, hairy on both sides *Source: Rocky Lemus*

Yellow hawkweed flowers *Source: James Altland*

Seed

- Size: 1.3—1.9 mm long
- Shape and texture: oblong, long ridges; smooth, has white to cream colored pappus
- Color: reddish brown to black

Field hawkweed plant *Source: Rocky Lemus*

Leaf shape and arrangement—Leaves: Narrow, elongated leaves 10—15 cm long; larger basal leaves and smaller, alternate stem leaves; the leaves are attached near ground level; the flowering stalk, arising from the leaf cluster, is leafless; leaves are hairy on both sides; nearly sessile.

Stolons/Rhizomes/Roots: Has stolons and rhizomes; the stems have short, stiff hairs and contain a milky latex sap inside the stem; the hairy stolons root at the node producing new rosette of leaves; fibrous root systems.

Inflorescence: The flowering stalks grow from a few cm to 90 cm in height. The flowers, which are about the same shade of yellow as buttercups, occur in heads that are about 2 cm in diameter.

Hawkweed, field—*Hieracium* L. spp.

Other common names: Meadow hawkweed, yellow hawkweed, yellow King Devil
Family: Compositae/Asteraceae
Life cycle: Cool-season perennial
Native to: Europe

Distribution and Adaptation
- Found throughout the southeastern United States.
- Commonly found in low maintained turf, pastures, roadsides, abandoned fields, permanent meadows, and range-lands; not usually found in cultivated fields.
- Well adapted in low fertility, shallow, dry, acid soils.

Morphology/Growth Pattern
- This stoloniferous and rhizomatous perennial grows in patches or alone; all leaves are basal with a few smaller leaves on the flowering stem; leaves are broadly club-shaped, smooth margins and hairy on both sides of the leaf surfaces; the flower heads are in branching terminal clusters at the top of the stalk; numerous florets, flowers yellow in color.
- Grows up to 1−3 ft (0.30−0.90 m) tall.
- Reproduces by seeds, stolons, and rhizomes.

Use and Potential Problems
- Unpalatable to livestock.
- If not controlled, can become invasive in pasturelands as well as crop fields.

Toxicity
- None recorded.

Similar Species
- Field hawkweed can easily be confused with *orange hawkweed* (*Hieracium aurantiacum* L.) however, orange hawkweed has bright orange flowers rather than yellow-like field hawkweed.

Reference
- Field hawkweed [33,46,47,58,72,76]

Healall—*Prunella vulgaris* L.

Healall seeds

Seed
- Size: 1.6—2.2 mm long, 0.9—1.1 mm wide
- Shape and texture: slightly flattened, pear-shaped if viewed lengthwise; has two flattened sides and third side that is rounded; smooth and shiny surface
- Color: brown to reddish-brown with dark vertical lines

Healall plant *Source: Lachlan Cranswick*

Healall leaves and square stem *Source: Lachlan Cranswick*

Leaf shape and arrangement—Leaves: Lance-shaped, broader at the base than at the tips, opposite, leaf margins are entire or shallowly and irregularly toothed/serrated and sometimes leaves are purple-tinged; may not have petioles on the upper leaves (sessile); young leaves and petioles are covered with hairs but usually no hair on older leaves.

Stolon/Rhizome/Roots: Stolons, no rhizomes; square stem, covered with hairs when young; roots are fibrous, shallow, and formed at the stem nodes.

Inflorescence: Flowering branches from stem at the axil; flowers are produced on erect, club-like spikes; immediately below this club-like head is a pair of green, hairy, stalkless bract-like leaves standing on either side, like a collar; flower colors range from light to dark purple, flowers have two lips and are tubular; the top lip is often purplish in color while the lower lip is often white.

Healall flower *Source: Lachlan Cranswick*

Healall—*Prunella vulgaris* L.

Other common names: Prunella, all-heal, hook-heal, self-heal, slough-heal, brunella, heart of the earth, blue curls, carpenter-weed, common selfheal, consolida minor, lance selfheal, sicklewort, woundwort, thimble-flower, thimble-weed, wild sage
Family: Lamiaceae
Life cycle: Perennial
Native to: Europe and North America

Distribution and Adaptation
- Found throughout the United States, common in the northeastern United States and Canada.
- Found growing in waste ground, grassland, woodland edges, usually on basic and neutral soils.
- Prefers moist conditions but can be found in a variety of sites ranging from dry to swampy, open to shady and cultivated to undisturbed sites.

Morphology/Growth Pattern
- A perennial, branched with upright to reclining growth habit; stems up to 2 ft (0.60 m) tall; has a variable growth habit, in moist wooded areas or crowded areas, healall grows tall, upright, and slender, whereas in lawns and frequently mowed or cut areas it spreads horizontally into dense thick patches; leaves elliptical to lance-shaped, broad at the base and tapered to a rounded tip, are numerous, opposite on a square stem; lower leaves with petiole and upper leaves sessile; leaves and stems are sparsely covered with hairs; flowers pale violet to purple, trumpet-shaped or tubular flowers appearing at the end of branches.
- Similar characteristics with others in mint family include square stem and opposite leaves; however, unlike many other members of the mint family, healall does not have the characteristic mint odor even when crushed, ovate-oblong leaves, relatively long leaf stalks (petioles), and roots emerging at stem nodes.
- Reproduces by seed and creeping stems that root at the nodes.

Use and Potential Problems
- No livestock feed value.
- Primarily weed of waste places, lawns, open wooded areas, pastures, and landscapes.

Toxicity
- None recorded.

Similar Species
- *Deadnettle, Purple or Red—Lamium purpureum* L. and *henbit* (*Lamium purpureum* and *Lamium amplexicaule* L.) are winter annuals that flower in early spring. Vegetative growth of red deadnettle can be distinguished from healall by its reddish stems and triangular leaves. Henbit is easy to distinguish from healall by its upper leaves that, lacking petioles, encircle the stem.

Reference
- Healall [14,18,33,39,47,67,72,76]

Henbit—*Lamium amplexicaule* L.

Henbit seeds

Seed

- Size: 1.5—2 mm long, 0.9—1.0 mm wide
- Shape and texture: long ovate, three-angled, ridges along the angles between the faces; surface smooth, dull, under high magnification fine spots/wrinkles are visible
- Color: dark gray-brown with many distinct white mottles that are more abundant around the apex; ridges are yellow

Purple deadnettle (left) henbit (right) leaves *Source: NCSU*

Henbit plant *Source: NCSU*

Henbit leaves, stem, and flowers

Leaf shape and arrangement—Leaves: Egg-shaped, hairy, with rounded teeth margins and prominent veins on underside; opposite, broad lower leaves with petioles; the upper leaves without petioles and clasping the stem.

Stolons/Rhizomes/Roots: No stolons or rhizomes; has a square stem; fibrous root system.

Inflorescence: Flowers reddish-purple with darker coloring in spots on lower petal, arranged in whorls in the axils of upper leaves; the flowers are two-lipped (with the upper lip being concave) that join to form a tube and four protruding stamens.

Henbit—*Lamium amplexicaule* L.

Other common names: Dead nettle, blind nettle, bee nettle, henbit deadnettle
Family: Lamiaceae
Life cycle: Winter annual
Native to: Europe and North Africa

Distribution and Adaptation
- Found throughout most of North America, also in West Indies, South America, Africa, Asia, and Australia.
- Thrives in cool, moist areas, and more commonly found in newly established fields and pastures.

Morphology/Growth Pattern
- Winter annual with square stems and pink-purple flowers, reaching 1.3 ft (0.40 m) in height.
- The plant primarily has an upright growth habit but can root from the lower nodes.
- Reproduces by seed and rooting stems.

Use and Potential Problems
- Primarily a weed of turfgrass, landscapes, pastures, and small grains.

Toxicity
- None recorded.

Similar Species
- Henbit is similar in appearance to *purple (red) deadnettle* (*Lamium purpureum* L.); the leaves of purple deadnettle, however, are more pointed at the end, are triangular, and less deeply lobed, and their lower leaves are on long petioles and the upper leaves are on short petioles; while only the lower leaves of henbit are petiolated. Additionally, purple deadnettle leaves are distinctly red or purple-tinted.

Reference
- Henbit [14,24,44,48,58,66,67,72,76]

Horsenettle, Carolina—*Solanum carolinense* L.

Horsenettle seeds

Horsenettle flower

Horsenettle flower

Seed

- Size: 2–2.4 mm long, 1.5–2 mm wide
- Shape and texture: ovate but variable to flat, round, whole seed may bent or curved, shallow notch along one edge near the smaller end of the seed; surface, smooth, glossy
- Color: yellow to orange-brown

Horsenettle plant

Leaf shape and arrangement—Leaves: Thick, alternate on spiny stem, leaf margins wavy to coarsely lobed edges; leaves have yellowish prickles (spines) located on their undersides along the midrib, leaves covered with small star-shaped hairs with four to eight spreading rays.

Stolons/Rhizomes/Roots: No stolons, deep rooted rhizomes with taproots; the stems have spines; taprooted.

Inflorescence: Flowers range in color from white to purple and have five petals; large, smooth, round tomato-like fruits are orange to yellow in color when mature (green when immature).

Horsenettle, Carolina—*Solanum carolinense* L.

Other common names: Horse nettle, bull nettle, apple-of-Sodom, wild tomato, sand brier, devil's potato, Carolina nightshade, tread-soft
Family: Solanaceae
Life cycle: Warm-season perennial
Native to: Southeast United States

Distribution and Adaptation
- Widely dispersed across the continental United States and Alaska, Hawaii, Puerto Rico, and the Virgin Islands.
- Prefers disturbed sandy or gravelly soils and waste areas.
- Will grow in a variety of soil types, but does best in sandy soils.
- Found in fields, pastures, woodlands, and waste areas especially in sandy soils.

Morphology/Growth Pattern
- Perennial herb with creeping rhizomes; may reach up to 3 ft (0.90 m) in height.
- Without the spines, the leaves would look very much like potato leaves.
- Reproduces by seed and rhizomes.

Use and Potential Problems
- Horsenettle can be a troublesome weed in pastures, orchards, landscape beds, and row crops.
- All parts of this species are toxic, it contains poisonous alkaloids including solanine. Should not be taken internally.
- The plant has been reported to be toxic to humans, cattle, sheep, and deer. However, massive consumption of this plant is rare due to the plants' prickly stems and leaves.

Toxicity
- Steroid alkaloids, glycosides, and aglycones.

Similar Species
- *Horsenettle* can be distinguished from the thistles because it has spines on its leaf midrib, while thistle spines are located on leaf margins.

Reference
- Horsenettle [12,14,19,24,33,44,46,48,65,67,72,76]

Horseweed—*Conyza canadensis* (L.) Cronquist

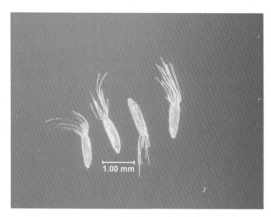

Horseweed seeds

Seed

- Size: 1–1.2 mm long, 0.3 mm wide (without pappus)
- Shape and texture: oblong but narrower toward the base, compressed with thin side margins, pappus of hairs extending from the apex 2–2.3 mm long, small collar around pappus; surface smooth, silky, many short white hairs
- Color: pale straw

Horseweed hairy stem and leaves

Horseweed plant

Horseweed flowers

Leaf shape and arrangement—Leaves: Narrowly lanceolate or oblanceolate, with a few teeth toward the outer tips, and fine white hairs along their margins; alternate, linear, entire, or toothed, crowded along the stem, upper leaves smaller than lower leaves on the stem, older leaves without *petioles*; seedling leaves form a basal rosette.

Stolons/Rhizomes/Roots: No stolons or rhizomes; stem erect, solid, and hairy, seldom branching below the inflorescence; short taproot, fibrous root systems.

Inflorescence: Numerous, small, inconspicuous flowering heads appear in terminal panicles; individual flowers are 5 mm in diameter with white or slightly pink ray flowers and yellow disk flowers accompanied by smooth green bracts that are somewhat longer.

Horseweed—*Conyza canadensis* (L.) Cronquist

Synonyms: *Erigeron canadensis* L., *Leption canadense* L.
Other common names: Mare's tail, fleabane, colt's-tail, Canada fleabane
Family: Compositae/Asteraceae
Life cycle: Summer annual
Native to: North America

Distribution and Adaptation
- Found throughout the United States, Canada, Central and South America.
- Prefers moist to dry, sunny conditions, and rich fertile soil.
- Drought resistant.

Morphology/Growth Pattern
- Tall and unbranched, except for the flower stems near the apex; can reach up to 6.5 ft (2 m) in height.
- Reproduces by seed.

Use and Potential Problems
- Weed of pastures, orchards, fallow fields, waste areas, roadsides, and native grassland.
- Less common in cultivated row crops.

Toxicity
- None recorded.

Similar Species
- At mature stage easily identifiable, however, at the rosette stage resembles other weeds with rosette growth habit such as *shepherd's-purse* (*Capsella bursa-pastoris* (L.) Medik.) and Virginia pepperweed (*Lepidium virginicum* L.).

Reference
- Horseweed [24,33,34,44,47,58,61,67,72,76]

Jewelweed—*Impatiens capensis* Meerb.

Impatiens spp. seeds

Jewelweed leaves

Jewelweed flower

Seed
- Capsule that pops open at maturity dispersing the seeds

Jewelweed plant

Leaf shape and arrangement—Leaves: Spade-shaped and rounded, slightly toothed; opposite on the bottom of the stem and alternate on the upper part; bright green, approximately 5−11 cm long and 3−7.6 cm wide; plant glabrous; leaves occur on petioles 6 cm in length.

Stolons/Rhizomes/Roots: No stolons or rhizomes; stems glabrous reaching up to 16.5 cm in height; flowering stem succulent, hollow, and juicy.

Inflorescence: Flower colored with varying shades of orange, spotted with a deeper red-orange, forming a triangular funnel with a spotted sepal at the "mouth" of the funnel; flower like a hanging basket.

Jewelweed—*Impatiens capensis* Meerb.

Other common names: Touch-me-not
Family: Balsaminaceae
Life cycle: Winter/summer annual
Native to: North America

Distribution and Adaptation
- Distributed throughout the United States.
- Found most often in damp areas, along the edges of streams and marshes, along roadsides and in noncultivated areas.
- Like most of its relatives, such as ornamental impatiens, jewelweed prefers shade to bright sunlight.

Morphology/Growth Pattern
- Annual herbaceous plant; it grows in a huge cluster, can reach 5 ft (1.5 m) in height; lower leaves opposite and upper leaves alternate; each leaf slightly toothed, thin, approximately 3.5 in. (8.8 cm) in length; stem translucent; the flowers are irregular in shape, have "wet," delicate appearance and are up to 1 in. (2.5 cm) long; flowers are orange and yellow with darker splotches; the seedpod looks like a hanging basket, the mature seedpod when touched or triggered by wind, pops open, hence the name touch-me-not.
- Plants can be easily identified by the translucent leaves, the uniform stand, and the characteristic hanging flowers.
- Reproduces by seed.

Use and Potential Problems
- Not consumed by livestock.
- Spreads rapidly, making it difficult to control.

Toxicity
- None recorded.

Similar Species
- Similar species include *pale touch-me-not* (*Impatiens pallida* L.); pale-touch-me-not has yellowish flowers and is larger than jewelweed.

Reference
- Jewelweed [24,58,67]

Jimsonweed—*Datura stramonium* L.

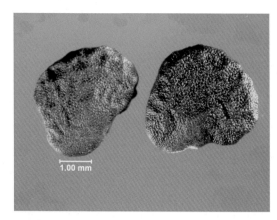

Jimsonweed seeds

Jimsonweed flower *Source: Nathan O'Berry*

Seed capsule covered with stiff prickles

Seed
- Size: 3−3.8 mm long, 2.5−3 mm wide
- Shape and texture: kidney-shaped, flattened, with pitted surface, slightly wrinkled, under high magnification in addition to the wrinkles, fine mesh-like patterns are visible
- Color: dull, dark brown to black

Jimsonweed plant

Leaf shape and arrangement—Leaves: Very angular, large, smooth (no hair), thin, wavy, coarsely toothed (jagged lobes) about 8−20 cm long, leaves on long stout petioles.

Stolons/Rhizomes/Roots: No stolons or rhizomes; thick, shallow, and extensively branched taproot system, stem stout, branched and green to purple in color.

Inflorescence: Flowers are large and trumpet or funnel-shaped (tubular), white to pinkish or lavender, borne singly on short stalks in the axils of branches, are attractive and fragrant; fruit is a spiny egg-shaped capsule covered with short, sharp spines; when the fruit is ripe the pods burst open splitting into four segments and scatter numerous poisonous black, kidney-shaped seeds.

Jimsonweed—*Datura stramonium* L.

Synonyms: *Datura tatula* L.
Other common names: Jamestown-weed, Jamestown lily, thorn-apple, stinkwort, stinkweed, mad-apple, trumpet plant, loco weed, angel's trumpet, devil's fireweed, dewtry, Apple of Peru
Family: Solanaceae
Life cycle: Summer annual
Native to: Asia

Distribution and Adaptation
- Found almost everywhere in the United States except in the North and West; most common in the South.
- Waste ground and cultivated land, prefers nitrogen-enriched habitats.

Morphology/Growth Pattern
- Herbaceous, annual plant that grows up to 3−5 ft (0.90−1.5 m) tall and even taller in rich soils.
- Dead leafless stem with dry seed remains standing in the field.
- Reproduces by seed.

Use and Potential Problems
- Primarily a weed of agronomic crops, but also found in disturbed areas, along roadsides, old fields, pastures, barnyards, hog lots, waste places, and gardens.
- Poisonous plant; all parts of the plant are toxic; however, the seeds, fruit, and leaves contain the highest level of alkaloids and are the usual source of poisoning in humans, cattle, goats, horses, poultry, sheep, and swine. Poisoning of humans in recent years is more frequent than livestock poisoning. Human poisoning results from sucking the nectar from flowers or consuming the seeds. Due to Jimsonweed's strong unpleasant odor and taste, animals avoid grazing it unless other more desirable forage species are not available.
- Alkaloids are related to those found in magic mushrooms; however, magic mushrooms do not cause death even if consumed in large quantities.

Toxicity
- The plant contains tropane alkaloids, which affect the central nervous system, with the major alkaloids being atropine and scopolamine.
- Symptoms associated with jimsonweed: blurred vision, confusion, agitation, and combative behavior.

Reference
- Jimsonweed [12,19,24,25,33,48,61,62,65−67,72]

Joepyeweed—*Eupatorium fistulosum* L., Previously Known as *Eupatorium purpureum* (L.) E.E. Lament

Joepyeweed opposite leaves and stem *Source: Wise Co. VA*

Leaf shape and arrangement—Leaves: Ovate to lanceolate, 6–18 cm long, 3–9 cm wide; often arching downward, margins finely serrated, glabrous, rarely pubescent; leaves on the stem close to the flower head are opposite, the leaves on the lower stem in whorls of five to six around the node.

Stolons/Rhizomes/Roots: No stolons, rhizomes; erect solid stems, 0.5–1.5 m tall; stems not usually spotted unlike *coastalplain joepyeweed* (*Eupatorium dubium* L.) which has a spotted stem with purple spots.

Inflorescence: Flowers are arranged in terminal clusters giving flat appearance; no more than eight tubular flowers; light pink in color.

Joepyeweed flower head *Source: Wise Co. VA*

Joepyeweed—*Eupatorium fistulosum* L. previously known as *Eupatorium purpureum* (L.) E.E. Lament

Other common names: Queen of the meadow, hollow joe-pye weed, trumpetweed
Family: Asteraceae
Life cycle: perennial
Native to: North America

Distribution and Adaptation
- Depending on the species, joepyeweed is adapted to both dry areas and fields as well as wet, sandy soils often near the cost from North Carolina to Maine.

Morphology/Growth Pattern
- Erect and robust perennial, with solid, purple spotted stems; dome-shaped flower heads can grow 6.5−10 ft (2−3 m) tall or taller.
- Foliage, when crushed, gives off vanilla smell.
- Reproduces by seed.

Use and Potential Problems
- No documented commercial values; joepyeweed flowers are attractive to butterflies such as the Monarch, and others.

Toxicity
- None recorded.

Similar Species
- *Coastal plain joepyeweed* (*E. dubium* L.) is similar to *sweetscented joepyeweed*. However, *coastal plain joepyeweed* is mostly found in the coastal regions and is only 1 m tall, has much wider leaves and the stems are spotted unlike sweetscented joepyeweed.

Reference
- Joepyeweed [24,46,58,76]

Knapweed, Spotted—*Centaurea stoebe* subsp. *micranthos* (Gugler) Hayek

Spotted knapweed seeds

Knapweed spp. leaves

Spotted knapweed *Source: John Wright*

Seed

- Size: 2.8−3.3 mm long, 1.2−1.6 mm wide
- Shape and texture: oblong with curved long edges, there is a large shallow notch on one long edge near the base; surface striate with 10−15 faint ridges, there are sparse fine white hairs
- Color: gray-brown, collar and ridges pale yellow

Knapweed spp. plant *Source: John Wright*

Leaf shape and arrangement—Leaves: Elliptical or oblong, generally with smooth margins, but can be slightly lobed; lower leaves are 5−10 cm long and deeply lobed; upper leaves are smaller, narrower; shoots and leaves are covered with dense gray hairs; no spine on leaves; very branched.

Stolons/Rhizomes/Roots: No stolons; extensive rhizomes; branching and hairy stems; has vertical and horizontal roots that have a brown to black, scaly appearance.

Inflorescence: Flowers are enclosed in solitary, flask-shaped heads on the ends of leafy branches, occur on shoot tips and generally are 0.5−1 cm in diameter; white to pink feathery flowers; have greenish to straw-colored bracts covered with overlapping pale bracts with papery margins.

Knapweed, Russian—*Centaurea repens* L.

Family: Compositae/Asteraceae
Other common names: Turkestan thistle
Life cycle: Warm-season perennial
Native to: Southern Russia and Asia

Distribution and Adaptation
- Found throughout the western United States; in Colorado, it grows on all types of soils and stands have been found to survive 75 years or longer.
- Grows on clay, sandy, or rocky prairies and sunny meadows, lakes, rivers, hills, and bottomlands.
- Grows well on saline soils.

Morphology/Growth Pattern
- Creeping perennial that reproduces from seed and vegetative root buds.
- Can grow to 3 ft (1 m) in height.
- Seeds remain viable for 2—3 years in soil.

Use and Potential Problems
- Weed forms dense, single species stands due to its allelopathic effects on other plants.
- Found in cultivated lands, grazing land, grain and other crops, waste places, roadsides, river banks, irrigation ditches, and waste places.
- Is difficult to control.

Toxicity
- None recorded.

Reference
- Russian knapweed [19,24,33,34,47,61,76]

Knotweed, Prostrate—*Polygonum aviculare* L.; Knotweed, Erect—*Polygonum erectum* L. or *Polygonum achorecum* L.

Prostrate knotweed seeds

Seed
- Size: 2.5–3 mm long, 1.5–2 mm wide
- Shape and texture: triangular to somewhat compressed, ovoid, pointed at apex; shiny along margins, striate to granular
- Color: reddish-brown to dark brown

Prostrate knotweed plant *Source: NCSU*

Prostrate knotweed plant showing stem, ocrea, and leaves *Source: NCSU*

Leaf shape and arrangement—Leaves: Broadly elliptical to oval-shaped leaves, rounded or pointed at the tip, tapering at the base; simple, alternate, small; leaves with small *petiole* or sessile; thin, papery, membranous, white leaf sheath (ocrea) covers the stem at the leaf base, smooth leaf edges, leaves bluish green in color.

Stolons/Rhizomes/Roots: No stolons or rhizomes; usually trailing radially from the base of the plant (taproot); stem lays flat on the soil surface, branched, wiry, slightly grooved, smooth; slender taproot; usually not rooting from the node.

Inflorescence: Flowers very small, one to five flowers occur in clusters in leaf axils, no petals, the pink color on the flower are sepals.

Knotweed, prostrate—*Polygonum aviculare* L.

Family: Polygonaceae
Other common names: Common knotgrass, mattgrass, doorweed, pinkweed, birdgrass, stonegrass, knotweed, wire-weed, way-grass
Life cycle: Summer annual
Native to: Europe

Distribution and Adaptation
- Found throughout the United States and Canada.
- Commonly found in disturbed, infertile, and compacted soils; also found in pathways, abandoned lots, fields, road-sides, and sidewalks.
- Grows well on infertile, acidic, and moist soils.

Morphology/Growth Pattern
- Grows prostrate or semierect, plants form circular mat up to 2.5 ft (0.76 m) in diameter, leaves are linear to elliptical.
- Easily identified by the white, thin membranous sheath (ocrea) that covers the stem at each node.
- Ocrea is a structure that is present in all the Polygonaceae family.
- Reproduces by seed.

Use and Potential Problem
- No livestock feed values.
- The thick mat this plant produces can aggressively compete with many desirable plants in turf and lawns.

Toxicity
- None recorded.

Similar Species
- Knotweed can be easily confused with some of the *spurges* (*Euphorbia* spp.); however, the *Euphorbia* species do not have an ocrea and produce milky sap when cut or damaged unlike prostrate knotweed; prostrate knotweed can also be confused with two other *erect knotweeds*: *Polygonum erectum* L. or *Polygonum achorecum* L., which have rounded or broadly elliptical, light green leaves and flower-stalks that are longer than the leaf sheaths.

Reference
- Prostrate knotweed [14,19,24,33,34,47,58,61,66,67,72,76]

Lamb's-quarters, Common—*Chenopodium album* L.

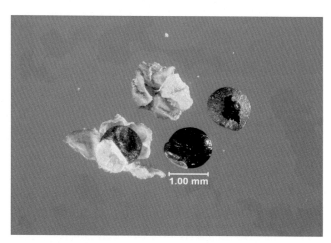

Common lamb's-quarters seeds; nick in one side of seed

Common lamb's-quarters leaves coarsely toothed *Source: James Altland*

Common lamb's-quarters stem *Source: James Altland*

Common lamb's-quarters seedhead; stem

Seed

- Size: 1.1–1.3 mm long, 1.0–1.1 mm wide
- Shape and texture: round with a nick in one side, lance shaped (one side flat, other convex); surface slightly roughened nearly smooth and shiny but not as shiny as redroot pigweed
- Color: black, unhulled seed may be enclosed in grayish or tan seed coat

Common lamb's-quarters plant

Leaf shape and arrangement—Leaves: Triangular in form with coarsely toothed lower leaves, upper leaves may be entire (no tooth) or indented margins; stalked, opposite, simple, upper surfaces green, but the undersides are whitened by powdery, waxy particles.

Stolon/Rhizome/Roots: No stolons or rhizomes; stem has prominent venation, erect, hairless, grooved, branching (many axillary stems at top of plant), sometimes has red to purplish color; root short, much branched taproot.

Inflorescence: Flowers small, inconspicuous, greenish, sessile, in irregular spikes clustered in panicles, borne at end of branches and in the axils of leaves; fruit covered by a thin papery pericarp.

Common lamb's-quarters flowers *Source: James Atland*

Lamb's-quarters, common—*Chenopodium album* L.

Family: Chenopodiaceae
Other common names: Lamb's quarters, pigweed, white goosefoot, fat-hen, mealweed, frostbite, bacon-weed
Life cycle: Summer annual
Native to: Native to Europe and Asia

Distribution and Adaptation
- Commonly found throughout the United States and much of Canada.
- Grows on both acidic and alkaline soils.

Morphology/Growth Pattern
- May grow up to 6 ft (1.8 m) depending on soil moisture and fertility.
- Succulent leaves and stems, and often has reddish-streaked stems.
- Reproduces by seed and is a very prolific seed-producer; may produce more than 70,000 per plant, however, many of these seeds are dormant and can stay viable in the soil for many years.

Use and Potential Problems
- Common weed of pastures, horticultural crops, cultivated crops, gardens, landscapes, waste places, barnyards, fence rows, and disturbed sites.
- Prior to the introduction of spinach, common lamb's-quarters was used as a wild vegetable, in salads, in soups and boiled. Considered a good source of vitamin A, calcium, potassium, protein, trace minerals, B-complex vitamins, vitamin C, and iron.

Toxicity
- **Common lamb's-quarters** contains oxalic acid as well as nitrates and is toxic to sheep and swine if eaten in large quantities over an extended period of time. However, this is not a common occurrence.

Reference
- Common lamb's-quarters [12,20,26,28,33,37,48,65,67,72]

Lettuce, Prickly Lettuce—*Lactuca serriola* L.

Prickly lettuce seeds

Prickly lettuce stem and clasping leaves

Prickly lettuce flower and seedhead

Seed

- Size: 2.7−3.0 mm long, 1.1−1.5 mm wide
- Shape and texture: oblanceolate with a small break, elongated, very thin narrow at the base and rounded at the apex; surface ribbed, dull, five to seven lengthwise ribs on each side; covered with symmetrical rows of hairs, has an appendage attached on one terminal
- Color: grayish brown; ribs and wings are pale yellow

Prickly lettuce leaves and stem

Leaf shape and arrangement—Leaves: Upper leaves are smaller and are more lanceolate than obovate, alternate on the stem, 5−30 cm long; the lower and basal leaves are sessile and have yellow prickly bristles on the midrib on the underside of the leaf; more prickly on the leaf margins, most leaves are deeply or shallowly lobed, the lobe tips are pointed backwards, leaf base clasps the stem; leaves bluish-green in color.

Stolon/Rhizome/Roots: No stolons or rhizomes; stem hollow, erect, stout, light green or yellowish; sometimes with stiff hair on stem; stem produces a milky sap when cut; taprooted.

Inflorescence: Pale to bright yellow flowers are produced on open, slender stalks, resembling cones with 13−24 ray flowers per head; the base of each flower head is enclosed by many greenish bracts; black to gray fruits, each with a long beak crowned with a tuft of white hairs at the tip.

Lettuce, prickly lettuce—*Lactuca serriola* L.

Synonymous: *Lactuca scariola* L.
Family: Compositate/Asteraceae
Other common names: Wild opium, wild lettuce, compass plant
Life cycle: Summer or winter annual
Native to: Europe

Distribution and Adaptation
- Distributed throughout the United States.
- Found in waste places, gardens, fencerows, pastures, roadsides, and overgrazed pastures; prickly lettuce thrives on nutrient-rich soils.

Morphology/Growth Pattern
- Sometimes referred to as wild lettuce because it is closely related to the cultivated lettuce.
- Erect, with few or no prickles, and grows up to 1−5 ft (0.3−1.5 m); the plant starts growing as a rosette and then a stem emerges from the central rosette, it then branches in the upper portion of the plant particularly in the inflorescence; the stem contains a milky sap; flowers are pale yellow become purple or bluish when dried; each flower head contains an average of 20 seeds with estimated seed production of up to 46,000 seeds per plant or more depending on the size of the plant.
- The key morphological characteristics of prickly lettuce include its lobed (sometimes not lobed), alternate, clasping leaves with prominent midrib which contains a row of spines on the underside of the leaf surface.
- Reproduces by seed and may also overwinter as a rosette.

Use and Potential Problems
- Prickly lettuce is a common weed in overgrazed pastures, horticultural crops, agronomic crops, also know to invade old alfalfa stands and fall planted alfalfa, and serious weed in no-till soybeans and winter wheat; prickly lettuce is drought tolerant so it can compete with grains for moisture during dry seasons.
- Is palatable to livestock and deer when young.
- Can be poisonous to livestock if large quantity is consumed.
- The sticky sap of prickly lettuce can clog harvesting equipment and can raise the moisture content of the grain.

Toxicity
- Lactones, lactucin, lactucopicrin.

Similar Species
- Prickly lettuce can be confused with *annual sowthistle* (*Sonchus oleraceus* L.); however, prickly lettuce is distinguished from the sowthistles by its prickled leaves along the margins and the midrib underside of the leaf surface, and unlike the sowthistles all parts of prickly lettuce produce milky sap.

Reference
- Prickly lettuce[12,19,24,33,47,66,76]

Mallow, Common—*Malva neglecta* Wallr.

Malva spp. seeds

Common mallow leaves, flowers, and fruits

Common mallow flower

Seed
- Size: 1.6—1.8 long, 1.5—1.7 mm wide
- Shape and texture: kidney shaped to nearly round with the notch in the inner angle, flattened lengthwise; surface finely textured, dull
- Color: light reddish-brown

Common mallow plant *Source: Fred Fishel*

Leaf shape and arrangement—Leaves: Kidney- to heart-shaped, and crinkled in appearance, about 2—6 cm wide, blunt to sharp toothed, simple to sometime lobed; alternate, on long *petioles* that are much longer than the leaf blade.

Stolon/Rhizome/Roots: No rhizomes or stolons; stems prostrate, freely branching with tip usually turned up and slightly hairy; stem lies close to the ground and can cover large areas; deep taproot with coarsely branched secondary root system.

Inflorescence: Flowers solitary or in clusters of one to three from leaf axils on a stalk (peduncle); white to light lavender with darker violet stripes on the petals, petals with notched tips; the numerous stamens are united by the filaments to form a cluster in the center of the flower.

Mallow, common—*Malva neglecta* Wallr.

Family: Malvaceae
Other common names: Round-leaved mallow, running mallow, malice, round dock, low mallow, button weed, dwarf mallow, cheeseweed, melba weed, mallow, malva
Life cycle: Winter annual, biennial, or perennial
Native to: Europe, Eurasia, and North Africa

Distribution and Adaptation
- Found throughout the United States in waste areas, disturbed habitats, gardens, cultivated land, roadsides, barnyards, and abandoned lots; more common in turfgrass, landscape, and nursery crops.

Morphology/Growth Pattern
- Creeping annual to short-lived perennial with prostrate to semierect, spreading, gray-brown, wrinkled stems that are 2–20 in. (5–50 cm) long; leaves with five to seven shallowly toothed lobes are long-stalked (petioled), rounded with heart-shaped base; flowers are white to pale.
- Due to its distinct characteristics, few plants are confused with common mallow.
- Reproduces by seed.

Use and Potential Problems
- Considered weed in pastures, lawns, turf, landscape, and gardens.
- If grown in soil highly fertilized with nitrogen, the plant can accumulate nitrate to a dangerous level.

Toxicity
- None recorded.

Reference
- Common mallow [24,47,66,72]

Mallow, Venice—*Hibiscus trionum* L.

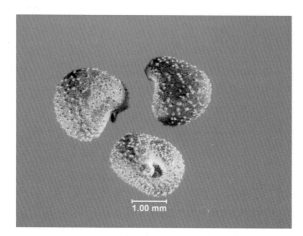

Venice mallow seeds

Venice mallow stem and leaves

Venice mallow flowers

Seed
- Size: 2–2.5 mm long, 1.6–2 mm long
- Shape and texture: mitt shaped, a large shallow notch in one side, there is a depression in the center of each face, smaller, more rounded and more pitted than velvetleaf; not as rounded as white campion; surface rough
- Color: dark brown to black

Venice mallow plant *Source: Fred Fishel*

Leaf shape and arrangements—Leaves: The first true leaves alternate and are irregularly shaped with toothed margins, the subsequent leaves are divided into three to five lobes, all leaves on long *petioles* and are glabrous (no hair) on the upper surface but with hairs (pubescent) on the lower surface.

Stolon/Rhizome/Roots: No stolons or rhizomes; stem herbaceous, erect, branching from the base, hollow, pubescent, thin vertical lines of hair in the upper stem, lines/nerves appear purple dotted because of the purple swollen bases of the hairs on them; shallow taproot and a fibrous root system.

Inflorescence: Flowers: solitary or in clusters, axillary; each flower has five sepals and five petals, numerous stamens form a cylinder around the pistil, clear creamy-white in color with deep purple bases; peduncles carry the flower up to or more than 2.5 cm long, longer when carrying fruit; fruit capsule inflated, hairy, dark-veined, and bell-shaped.

Mallow, Venice—*Hibiscus trionum* L.

Family: Malvaceae
Other common names: Venice mallow, flower of-an-Hour, good-night-at-noon, bladder ketmia, modesty, shoo-fly
Life cycle: Summer annual
Native to: Europe

Distribution and Adaptation
- Found throughout the United States but most common in the eastern half of the United States.
- Habitat includes pastures, crop fields, and waste places.

Morphology/Growth Pattern
- A short stemmed, erect or spreading, branching at the base from a taproot, hairy, annual plant; can grow up to 10−18 in. (18−46 cm) in height; the lower stems and petioles can be red to purplish in color; leaves alternate, long stalked (petioled), palmately divided, three to five coarsely toothed lobes; flowers are showy, with five petals and five sepals, petals are white to yellowish white, with reddish purple bases.
- Reproduces by seed.

Use and Potential Problems
- Venice mallow can be a problematic weed in agronomic crops, nursery crops, and pastures.
- Seeds can remain in the soil dormant up to 50 years, thus remove plants prior to seed development.

Toxicity
- None recorded.

Reference
- Venice mallow [19,24,47,58,66,72]

Milkweed, Common—*Asclepias syriaca* L.

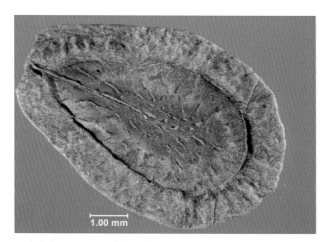

Milkweed seed

Common milkweed flower

Common milkweed fruit

Seed
- Size: 8–10 mm long, 4.4–5.5 mm wide
- Shape and texture: pear shaped, flattened, with a truncate small end and a thin marginal wing; surface papery, dull, many fine irregular wrinkles
- Color: tan or light brown margins with darker reddish-brown central part

Milkweed plant

Leaf shape and arrangement—Leaves: Widely elliptic to ovate-elliptic in shape and 10–25 cm long and 3–10 cm wide; are opposite, large (up to 25 cm) with entire margins, prominent white mid-veins on leaves with minute hairs that give them a fuzzy look.

Stolon/Rhizome/Roots: No stolons or rhizomes, stems are reddish and produce milky sap when injured or broken; has an underground creeping, strong root-stock, with extensive horizontal and vertical root systems.

Inflorescence: Flowers are in a rounded umbel; each flower is five-parted, pinkish-purple in color which when mature droops down; fruits green and turn brown at maturity, and bursts open to reveal a fluffy hair mass; the seedpod is filled with tiny seeds each with a tuft of silky hairs which become airborne.

Milkweed, common—*Asclepias syriaca* L.

Family: Asclepiadaceae
Other common names: Silkweed, milkweed, milkweed root, silkweed, silky swallow wort, cottonweed, Virginia silk
Life cycle: Warm-season perennial
Native to: North America

Distribution and Adaptation
- In the United States found throughout the eastern half of the United States except Louisiana and Texas.
- Adapted to all types of soil textural classes, but thrives on well-drained, loamy soils.
- Found in fields, gardens, pastures, open woods, roadsides, railroads, fencerows, and waste ground.

Morphology/Growth Pattern
- An upright, unbranched, large, broad leaved, perennial herb; the leaves are opposite, simple, short-stalked (on short petiole); broad leaves with prominent white mid-veins; leaves green on the upper surface but light-colored and velvety-hairy on the lower surface; has a strong root stock that grows horizontally to pop up clones nearby and may form a colony; stem simple usually single; plant produces a milky sap/juice when injured; flowers are in cluster at the top of the stem or in the axils of upper leaves, five-parted, flower color pinkish-purple; seedpods are 3–4 in. (8–10 cm) long, covered with soft spines and soft hairs and are inflated; when dry burst open producing many hair-tufted seeds.
- Grows up to 2–5 ft (0.6–1.5 m) tall.
- Reproduces by seed, and by adventitious buds found on horizontal root stock and stem base/crown near the soil surface.

Use and Potential Problems
- Considered by many as troublesome, while others consider the plant to be good for wildlife especially butterflies.
- A problem weed mostly in corn, soybeans, peanuts, and grain sorghums.
- Unless forage is in a short supply, animals avoid grazing milkweed.

Toxicity
- The milky sap produced by the leaves and stem is poisonous to humans, as well as all livestock, although sheep are at most risk, other animals like cattle, horses, goats, poultry, and pets are equally affected.

Similar Species
- Where found, common milkweed can easily be confused with *purple milkweed* (*Asclepias purpurascens* L.) however, purple milkweed has bright purple flowers located at the apex of the plant and also are mostly found in swamp forests; another plant often confused with common milkweed is *hemp dogbane* (*Apocynum cannabinum* L.), the similarities between these two species is that they both produce milky sap, hemp dogbane can be easily distinguished from common milkweed by its narrow, nearly sessile, elliptical leaves 2–5 in. (5–12 cm) long, and the upper one-third of the stem is much branched; flowers of hemp dogbane are small, greenish-white, and bell-shaped; unlike common milkweed, hemp dogbane has a characteristic, long, narrow, and curved fruit that dangles down the plant.

Reference
- Common milkweed [1,6,12,19,24,33,34,39,46,58,61,66,67,72,76]

Morningglory, Tall—*Ipomoea purpurea* (L.) Roth

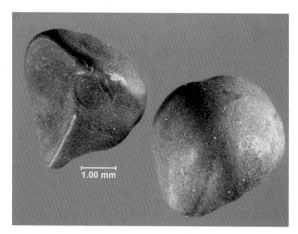

Tall morningglory seeds

Tall morningglory leaf *Source: Nathan O'Berry*

Tall morning glory flower *Source: Nathan O'Berry*

Seed
- Size: 4.7−5.7 mm long, 3.3−4 mm wide
- Shape and texture: orange slice shaped with horseshoe hilum; dorsal side strongly convex with shallow, central longitudinal furrow; ventral side has two equal, flattened faces with one or two cross wrinkles and thin lines along the margins
- Color: dark colored with dark hilum.
- Seeds per pound: 12,800

Tall morningglory plant *Source: Nathan O'Berry*

Leaf shape and arrangement—Leaves: Heart-shaped leaves, alternate, simple, long-*petioled*, older leaves have rounded base that overlap, leaf pointed at end; hairs lie flat against the leaf surface.

Stolon/Rhizome/Roots: No stolons or rhizomes; stems hairy, trailing, or climbing, capable of reaching 17 cm in length.

Inflorescence: Flowers occur in clusters of three or more and varies in color from purple to blue, white and pink, pubescent, flowers are funnel-shaped with short, broad sepals (green appendages at the base of the flower); fruit globular capsule containing four to six dark brown to black seeds.

Morningglory, tall—*Ipomoea purpurea* (L.) Roth
Morningglory, ivyleaf—*Ipomoea hederacea* Jacq.
Morningglory, pitted—*Ipomoea lacunosa* L.

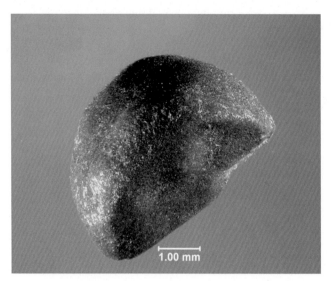

Ivyleaf morningglory

Ivyleaf morningglory *Source: Peter Sforza*

Cutleaf morningglory leaves and stem *Source: Nathan O'Berry*

Ivyleaf morningglory plant *Source: Fred Fishel*

Morningglory, tall—*Ipomoea purpurea* (L.) Roth.

Family: Convolvulaceae
Other common names: Common morningglory, wild morningglory

Morningglory, ivyleaf—*Ipomoea hederacea* Jacq.

Other common names: Wild morningglory, annual morningglory

Morningglory, pitted—*Ipomoea lacunosa* L.

Other common names:
Life cycle: Summer annuals
Native to: Tropical America

Distribution and Adaptation
- Prefers well-drained, organically rich soil.
- Found throughout the eastern half of the United States except for the far northern states.

Morphology/Growth Pattern
- A trailing or climbing annual vine with heart-shaped leaves and purple to white flowers.
- Reproduces by seed.

Use and Potential Problems
- Primarily a weed of agronomic crops, nurseries, landscapes, and some pastures. Invades disturbed places, forest openings, and right-of-ways.

Toxicity
- Ergot-type neurotoxic indole alkaloids.

Similar Species
Tall morningglory is often confused with the other morningglories: *Pitted morningglory* (*Ipomoea lacunose* L.) and *Ivyleaf morningglory* (*Ipomoea hederacea* Jacq.). Pitted morningglory looks like tall morning glory except that pitted morningglory has smaller leaves, leaf tips are pointed and leaf is hairless. Ivyleaf morningglory leaves are alternate, deeply three-lobed, ivy-like, with rounded base and hairs on stem, petioles and leaves are erect. Generally, tall morningglory is easily distinguished from the other morningglories by its distinct heart-shaped leaves that often overlap one another at the base and purple to white flowers.

Reference
- Morningglory [12,18,33,46,48,61,65,67,72,76]

Motherwort—*Leonurus cardiaca* L.

Motherwort leaves

Motherwort plant *Source: Lachlan Cranswick*

Motherwort flower head *Source: Lachlan Cranswick*

Leaf shape and arrangement—Leaves: Lower leaves are divided (three-lobed with lobes sharply pointed) or palmately three to five lobed-shaped and have irregular, saw-toothed margins; the upper leaves much smaller, usually divided into three parts and have slightly toothed margins; opposite, long-stalked, blades 5—12 cm long and almost as wide; petioles 3—5 cm long; glabrous.

Stolon/Rhizome/Roots: No stolons, rhizomes; square stem and may have short hairs along the corner; shallow extensive root system.

Inflorescence: Flowers are arranged in dense whorls/clusters and consist of short tubes and two lips; produced in the axils of the leaves; sepals located directly beneath the flower have prominent spiny/prickly tips; the petals are pale-purple and bearded; seedpods are arranged in spiny clusters.

Motherwort—*Leonurus cardiaca* L.

Family: Labiatae
Other common names: Common motherwort, cowthort, lion's-ear, lion's-tail
Life cycle: Perennial
Native to: Europe and Asia

Distribution and Adaptation
- Introduced from Europe; in the United States.
- Found in wooded areas, most frequently in some shade.

Morphology/Growth Pattern
- The square stem plant has a stiff stem, slightly branched and nearly smooth; flowers pink or purple in color are produced in cluster of 6–15 at the axils of upper leaves attached to the stem; several stems can be produced from a single root crown; can grow up to 3–4 ft (0.90–1.2 m) tall.
- The plant like other members of its family has square stems and will release pungent odor if injured or crushed.
- Reproduces by seed and underground stems (rhizomes).

Use and Potential Problems
- Considered a weed.
- Found in pastures, meadows, fields, yards, abandoned fields, open woods, riverbanks, waste areas, and along road sides; species prefer moist and fertile soils.

Toxicity
Motherwort contains lemon-scented oil that causes photosensitivity if ingested; leaves also can irritate skin in people who are susceptible.

Reference
- Motherwort [19,33,36,47,58,67,76]

Mullein, Common—*Verbascum thapsus* L.;
Mullein, Moth—*Verbascum blattaria* L.

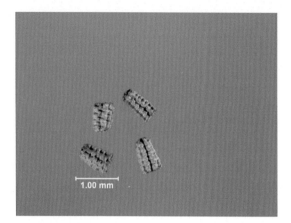

Common mullein seeds

Common mullein flower head

Common mullein flower head

Seed (Common Mullein)
- Size: 0.6–0.9 mm long, 0.4–0.5 mm wide
- Shape and texture: trapezoid shaped, wide at base, tapered to slightly rounded base; rough surface with 8–10 wavy length-wise ridges that are intersected by several irregular ridges forming oblong projections separated by deep, wavy vertical grooves
- Color: light brown with dark brown near base or entire seed dark brown

Common mullein plant

Leaf shape and arrangement—Leaves: Oblong, rosette leaf, leaves on the upright stem are alternate, extremely hairy with entire or bluntly toothed margins, upper leaves are sessile, narrower and more pointed at the apex than the basal leaves. Moth mullein basal leaves are sessile, whereas the common mullein leaves are on short petioles, moth mullein leaves are smooth (no hair) and have toothed margins, stem leaves are oblong and sharply pointed, alternate, clasping, and smooth.

Stolon/Rhizome/Roots: No rhizomes or stolons; slender, erect, unbranched stem; thick fleshy taproot with shallow secondary fibrous root system.

Inflorescence: Common mullein flowers in dense terminal spike, nearly sessile, individual flowers have yellow fused petals with five lobes, sepals have five lobes and are wooly; fruits are egg-shaped; moth mullein flowers are born on pedicels, yellow to white, with five fused petals along an elongated axis, with soft purple hairs on the stamens.

Mullein, moth—*Verbascum blattaria* L.

Moth mullein flowers *Source: Wise Co.*

Moth mullein stem and leaves *Source: Lachlan Cranswick*

Moth mullein flower *Source: Lachlan Cranswick*

Mullein, common—*Verbascum thapsus* L.

Family: Scrophulariaceae
Other common names: Velvet dock, big taper, candle-wick, flannel-leaf, woolly mullein
Life Cycle: Warm-season biennial

Mullein, moth—*Verbascum blattaria* L.

Other common names:
Life Cycle: Warm-season biennial
Native to: Europe

Distribution and Adaptation
- Found throughout the United States except for the upper great plains and southern Canada.
- Often found where the soil is dry and gravelly.

Morphology/Growth Patterns
- Moth mullein plant generally is less robust than common mullein and may reach 5 ft (1.5 m) in height.
- Reproduces by seed.

Use and Potential Problem
- Weed of landscapes, pasture, perennial crops, and roadsides.

Toxicity
- None recorded.

Reference
- Common mullein [19,24,33,61,66,67,72,76]

Mustard, Hedge—*Sisymbrium officinale* (L.) Scop.

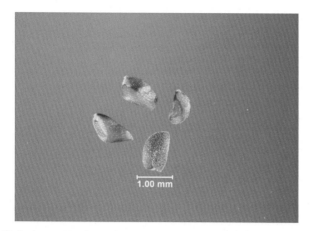

Hedge mustard seeds

Seed
- Size: 1.1−1.6 mm long, 0.5−0.7 mm wide
- Shape and texture: shape variable, from elliptical to rounded oblong or with one straight edge and one convex edge; surface dull, finely roughened, semiglossy
- Color: orange-brown

Hedge mustard plant

Hedge mustard leaf

Hedge mustard seedpod

Leaf shape and arrangement—Leaves: Alternate, deeply lobed, or divided; both side of the leaf surface is hairy; the uppermost part of the leaf is the largest and most distinctly toothed; lower leaves long-stalked (petioled) and deeply dissected, upper leaves are less dissected and are short-stalked.

Stolons/Rhizomes/Roots: No stolons or rhizomes, the plant produces a strong slender taproot 10 cm long or more even at the rosette stage.

Inflorescence: Pale yellow small flowers are short-stalked and arranged in clusters, the four-petaled flowers are 3−4 mm in diameter; narrow and sharp-pointed seedpods located close and parallel to the stem and are 10−20 mm long seedpod; each pod contains several hard brown seeds.

Hedge mustard flower

Mustard, hedge—*Sisymbrium officinale* (L.) Scop

Family: Brassicaceae
Synonyms: *Chamaeplium officinale* L., *Erysimum officinale* L.
Other common names: Wireweed, hedge weed
Life cycle: Winter or summer annual
Native to: Europe and Asia

Distribution and Adaptation
- In the United States found throughout most of the states; rare or absent in the prairies.
- Found in waste places, roadsides, cultivated fields, and gardens.
- Grows on droughty, fertile loamy, or sandy soils.

Morphology/Growth Pattern
- The annual plant briefly forms a rosette and quickly branches; the erect plant, has a wiry stem and reaches up to 3 ft (0.90 m) in height.
- Hedge mustard initially appears like dandelion, however, unlike dandelion, the indentation or tooth on the leaf of hedge mustard goes to the stem with slight gap between each leaf; also the stem turns purple.
- Reproduces by seed.

Use and Potential Problems
- Hedge mustard can cause a significant loss particularly in mustard crops such as canola, turnips, or cabbage.
- Animals avoid grazing hedge mustard.

Toxicity
- Glucosinolates

Reference
- Hedge mustard [12,24,33,47,72,76]

Mustard, Pinnatetansy—*Descurainia brachycarpa* L.

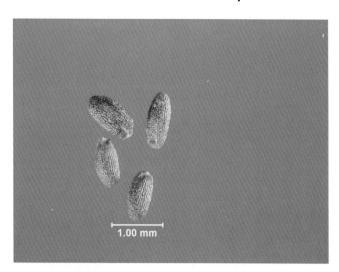

Pinnate tansymustard seeds

Seed

- Size: 0.9—1.3 mm long, 0.4—0.6 mm wide
- Shape and texture: oblong, with one convex and one nearly flat face, some seeds have thin transparent marginal wing on the rounded end, opposite the notch; surface smooth
- Color: dull orange

Pinnate tansymustard *Source: Wise Co.*

Pinnate tansymustard plant *Source: Wise Co, VA*

Leaf shape and arrangement—Leaves: Long finely divided almost fern-like leaves; basal leaves are stalked and divided twice in small segments, fine, dense star-shaped hairs, 3—10 cm long, the leaves on the stem are divided into small segments once and are densely hairy.

Stolons/Rhizomes/Roots: No stolons or rhizomes, one or several densely hairy stems.

Inflorescence: Flowers are small, bright yellow to almost white produced on elongated racemes; seedpod narrow, smooth, slightly curved, two-celled, many seeded.

Mustard, pinnatetansy—*Descurainia brachycarpa* L.

Family: Brassicaceae
Synonyms: *Descurainia sophia* (L.) Webb. Ex Prantl
Other common names: Crownbeard, flixweed, western tansymustard
Life cycle: Cool-season, winter annual
Native to: North America

Distribution and Adaptation
- Found throughout the United States and Canada.
- Grows well in dry, sandy, and saline soils; found in open or sparsely wooded places, waste places, fields, roadsides, and disturbed sites.

Morphology/Growth Pattern
- Winter annual plant that can grow 4–32 in. (10–81 cm) tall; leaves alternate on the stem, pinnately dissected (looks like fern), hairy; the many clustered flowers are tiny, pale yellowish to whitish; flower has both male and female organs in the same plant; long seedpod each having two rows of seed.
- Reproduces by seed.

Use and Potential Problems
- Weed in turf, in row crops and pastures.

Toxicity
- None recorded.

Similar Species
Pinnated tansymustard and flixweed (*Descurainia sophia L.*) are very similar and often confused with one another. They both grow as a rosette with finely lobed compound leaves, both produce yellow flowers, and small orange seeds are produced in long, narrow seedpods.

Reference
- Pinnate tansymustard [14,44,74,76]

Mustard, Wild—*Brassica nigra* (L.) W.D.J. Koch

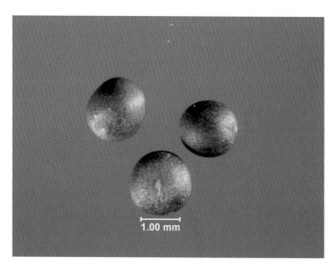

Wild mustard seeds

Seed
- Size: 1.5—1.7 mm long, 1.4—1.6 mm wide
- Shape and texture: round or broadly elliptical; surface smooth, dull
- Color: black to reddish brown, the mesh on the surface is whitish

Wild mustard *Source: James Altland*

Wild mustard flower with four yellow petals *Source: James Altland*

Leaf shape and arrangement—Leaves: Lanceolate and not pinnately lobed and are slightly toothed; alternate on the stem; leaves are prominently veined; some hair on the underside of the leaves; leaves are broadest at the tip and tapers at the base; lower leaves are petioled and have deep, irregularly lobed, upper leaves with or without petiole, become progressively smaller.

Stolons/Rhizomes/Roots: No stolons or rhizomes; stems are erect and branched; slender taproot with branching secondary roots.

Inflorescence: Flowers are yellow in color and contain four green sepals and four yellow petals 8—12 mm long; long characteristic seedpod which is distinctly ribbed with knob on end, each seedpod contains 10—20 seeds.

Mustard, wild—*Brassica nigra* (L.) W.D.J. Koch

Family: Brassicaceae
Synonyms: *Brassica arvensis* L., *Sinapis arvensis* L.
Other common names: Crownbeard, charlock mustard, field mustard, field kale, kedlock, common mustard, kaber mustard, water cress, yellow-flower, Herrick, yellow mustard
Life cycle: Winter or sometime summer annual
Native to: Europe

Distribution and Adaptation
- Found throughout the United States.
- Infests roadsides, cultivated fields, ditch banks, waste areas, gardens, along fence rows, and sometime in pastures.

Morphology/Growth Pattern
- The flowering stem of wild mustard is erect and can reach 8–32 in. (20–81 cm) tall; leaves are hairy and alternately arranged on the stem, 1–3 in. (2.5–8 cm) wide and 2–8 in. (5–20 cm) long with irregular or toothed margins; four-petaled yellow flowers; distinct, long slender seedpod, tipped with a small knob.
- Reproduces by seed and seed remains in soil for many years.

Use and Potential Problems
- Wild mustard is mostly considered as a weed with no commercial value.
- An aggressive competitor with crops specially with canola; besides competing with canola for soil moisture and nutrients, the seed in canola can lower the quality of the canola oil.
- Common weed of field crops and pastures as well as nurseries, waste areas, and disturbed sites.

Toxicity
- Glucosinolates.

Similar Species
Yellow rocket (*Barbarea vulgaris* L.) and *wild radish* (*Raphanus raphanistrum* L.) closely resembles wild mustard. Generally wild mustard and yellow rocket are similar in growth habit and appearance; however, neither of these plants are covered with stiff hairs as wild radish. Also, the leaves of wild radish are more divided than wild mustard or yellow rocket.

Reference
- Wild mustard [12,33,36,61,66,67,72,74]

Mustard, Yellow Rocket—*Barbarea vulgaris* **R. Br.**

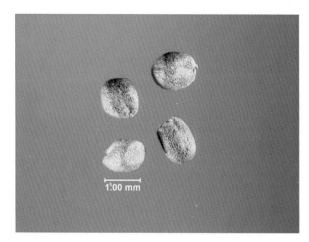

Yellow rocket seeds

Yellow mustard stem and leaves

Yellow rocket flower *Source: Lachlan Cranswick*

Seed
- Size: 1.4−1.7 mm long, 1.1−1.3 mm wide
- Shape and texture: oblong with a notch in one short side; surface finely textured, slightly sheen, at a high magnification, a fine honeycomb mesh is apparent
- Color: brown to gray

Yellow rocket plant

Leaf shape and arrangement—Leaves: In the first year a rosette of leaves formed on the ground surface; leaves alternate, hairless, basal leaf large, petioled and all or mostly pinnately compound to lobed, with large round terminal lobe, one to four small lateral lobes, the upper leaves are similar to the basal leaves but short-petioled to sessile and clasping, the uppermost leaves are shallowly lobed or toothed, become smaller and less toward the tip of the plant; 3−15 cm long including petiole; leaves are dark green in color.

Stolons/Rhizomes/Roots: No stolons or rhizomes; taprooted, with dense fibrous root system.

Inflorescence: Bright yellow flowers each with four petals, 5−8 mm long, sepals 2−3 mm long; slightly curved, slender seedpod, seedpod 10−30 mm long, 0.5−1 mm thick.

Mustard, yellow rocket—*Barbarea vulgaris* R. Br.

Family: Brassicaceae
Other common names: Crownbeard, winter cress
Life cycle: Cool-season, biennial or short lived perennial
Native to: Europe and Asia

Distribution and Adaptation
- Found throughout the United States and Canada.
- Found in wet meadows, stream banks, roadsides, and waste places.

Morphology/Growth Pattern
- The plant can grow up to 8–24 in. (20–60 cm) long; the lower part of the stem is simple, branched above; the base of the plant is often purplish; the first year a rosette of leaves is formed on the surface of the ground, followed by flowering stem the second year; the rosette leaves and leaves on the lower part of the stem are on long petiole and lobed with the terminal lobe being the largest.
- Reproduces by seed.

Use and Potential Problems
- No commercial value, knows as a weed.

Toxicity
- Glucosinolates.

Similar Species
- The flowers of yellow rocket are similar to several other mustard species such as wild mustard (*B. kaber*) and wild raddish (*Raphanus raphanistrum*). However, wild radish leaves are covered with hairs unlike the leaves of wild mustard or yellow rocket.

Reference
- Yellow rocket [12,33,39,47,61,66,72,76]

Nightshade, Bittersweet—*Solanum dulcamara* L.;
Nightshade, Eastern Black—*Solanum ptychanthum* Duanl;
Nightshade, Silverleaf—*Solanum elaeagnifolium* Cav.

Bittersweet nightshade seeds

Bittersweet nightshade leaf

Bittersweet nightshade flowers *Source: Rocky Lemus*

Seed
- Size: 2–2.6 mm long, 1.7–2 mm wide
- Shape and texture: kidney-shaped, flattened, indentation near hilum
- Color: pale yellow
- Pattern of fine veins on surface

Bittersweet nightshade plant

Leaf shape and arrangement—Leaves: Broadly ovate, entire margins, often with two basal lobes (ear-like lobes) or leaflets at the base (not always present), *petiolated*; alternate, simple, smooth entire margins; hairless, upper part of leaf dark green and slightly lighter below.

Stolon/Rhizome/Root: No stolons or rhizomes; a trailing or climbing woody vine; roots at the node; fibrous root system.

Inflorescence: Flowers appear in clusters, bright purple petals (occasionally white), yellow anthers; fruit: bright red, egg-shaped berries arranged in clusters.

Nightshade, bittersweet—*Solanum dulcamara* L.

Other common names: Bittersweet, nightshade, climbing nightshade, woody nightshade, violet-bloom, blue bindweed, poisonberry, poisonflower, morel, snakeberry, wolfgrape, scarlet berry, tether-devil
Family: Solanaceae
Life cycle: Cool-season perennial
Native to: Eurasia

Distribution and Adaptation
- Found throughout most of the United States, most common in the eastern and north-central states.
- Grows in sun or partial shade, dry to moist soil conditions.

Morphology/Growth Pattern
- Low climbing, scrambling, sprawling vine often draping low over trees and shrubs.
- Can grow to a height of 3.3−9.8 ft (1−3 m).
- Reproduces by seed.

Use and Potential Problem
- Weed in disturbed areas, roadsides, edge of moist woods, waste places, fencerows, forest or natural area.

Toxicity
- Steroid alkaloids, glycosides, and aglycones.

Similar Species
- Bittersweet nightshade is often confused with *Eastern black nightshade* (*Solanum ptycanthum* L.); however, eastern black nightshade is an annual with an upright growth habit and the leaves of this plant have wavy margins and the berries are black instead of red like bittersweet nightshade.

Reference
- Bittersweet nightshade [12,19,24,58,62,73,74,76]

Nightshade, Eastern Black—*Solanum ptychanthum* Duanl; Nightshade, Silverleaf—*Solanum elaeagnifolium* Cav.

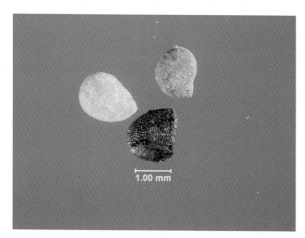

Eastern black nightshade seeds

Eastern black nightshade showing holes in the leaves *Source: Lachlan Cranswick*

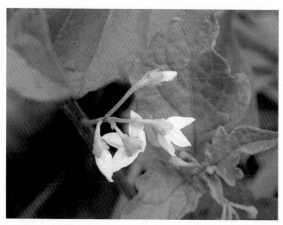

Black nightshade flowers *Source: James Altland*

Seed
- Size: 1.7–2.1 mm long, 1.4–1.6 mm wide, much smaller than horsenettle
- Shape and texture: asymmetrically obovate, with a blunt tip at the small end; surface smooth, dull under high magnification, appears finely pitted and minutely wrinkled
- Color: yellow to dark brown

Eastern black nightshade plant *Source: Lachlan Cranswick*

Leaf shape and arrangement—Leaves: *Eastern Black Nightshade*: ovate or ovate-lanceolate, shaped like tomato leaves, with hair or hairless; margins may be entire or with blunt teeth; simple, alternate; young leaves remain purple or maroon-tinted on the undersurface. *Silverleaf nightshade*: leaves usually armed with small sharp spines or thorns, entire or wavy-margined; silvery-white (gray underside) with a dense covering of star-shaped clusters of hair; often, holes are found in the leaves of both eastern black and silverleaf nightshade.

Stolon/Rhizome/Roots: No stolons or rhizomes; erect stem, widely branched, stems arise from deep-seated rootstock and creeping roots; a taproot with a branched fibrous root system.

Inflorescence: *Eastern black night shade* flowers are star-shaped, white or purple-tinged, in umbel-like clusters of five to seven; berry 5–12 mm in diameter, fruits are small green "tomatoes," turning deep purple at maturity. Silverleaf nightshade flowers are violet or blue with yellow or orange anthers.

Nightshade, eastern black—*Solanum ptychanthum* Duanl

Other common names: Deadly nightshade, poison berry, garden nightshade
Family: Solanaceae
Life Cycle: Warm-season annual

Nightshade, silverleaf—*Solanum elaeagnifolium* Cav.

Other common names: White horse nettle, bull nettle
Life cycle: Warm-season, short-lived perennial
Native to: Europe and the United States

Distribution and Adaptation
- Native to the mid-western states (silverleaf nightshade); eastern black nightshade native to Europe.
- Eastern black nightshade widespread from southern Canada throughout the United States to Mexico; most common east to the Rocky Mountains.

Morphology/Growth Pattern
- Eastern black nightshade is a highly branched annual plant that grows 1–3 ft (0.30–0.90 m) tall.
- The purple-red color usually present on the lower surface of the seedling leaves of eastern black nightshade helps distinguish it from other nightshade species.
- Flea beetles are known to feed on the leaves of nightshades leaving many distinct holes in the leaves.
- Reproduces by seed.

Use and Potential Problems
- Primarily a weed of cultivated fields, hay fields, gardens, waste places, and overgrazed pastures.
- Causes serious economic losses to soybean producers.

Toxicity
- Steroid alkaloids, glycosides, and aglycones.

Reference
- Eastern black nightshade [12,24,25,33,47,48,58,61,65,67,72,74,76]

Onion, Wild—*Allium* L. spp.

Wild onion seed

Wild onion plant

Wild onion flower head

Seed
- Shape and texture: smooth
- Color: actual seed is black

Wild onion stem pulbets

Leaf shape and arrangement—Leaves: Several long grass-like leaves are produced from papery sheathing bases which overlaps and close the stem that arises from the bulb and a few inches above ground; mature leaves are brown, rolled and reduced to the diameter of string.

Stolon/Rhizome/Roots: No stolons or rhizomes; stem slender; long, smooth, and hollow.

Inflorescence: Compact flower heads white or pink in color appear in a compact, round umbel one at the tip of each slender stem; the umbel is enclosed by two or three papery bracts before flowers open; flowers have three petals and three sepals that are all petal-like; at the end of flowers replaced by bulblets; are first onion-like but become flattened; one bulb per plant covered with a purplish or reddish coat

Onion, wild—*Allium* L. spp.

Other common names: Meadow garlic, crow garlic, field garlic, scallions
Family: Liliaceae
Life cycle: Cool-season, perennial
Native to: Europe

Distribution and Adaptation
- Commonly found in cultivated crops, pastures, and meadows.

Morphology/Growth Pattern
- The stem and leaves are smooth, long narrow hollow and have a dark green color; leaves originate from the base of the plant usually at the ground level; commonly grows 6–18 in. (15–46 cm) tall.
- Reproduces by bulbs, bulblets, and sometimes by seed.

Use and Potential Problem
- Found in pastures and meadows; dairy animals on pastures grazing forages with wild garlic in it produce milk with an onion or garlic taste.

Toxicity
- None recorded.

Similar Species
- *Wild onion* (*Allium canadense* L.) is often confused with *wild garlic* (*Allium vineale* L.). Both plants often occupy the same sites. Wild onion can be distinguished from *wild garlic* by its leaves that are flat in cross section and not hollow; the presence of a fibrous coat on the central bulb, the offset bulblets at the base of the plant and the areal bulblets readily distinguish wild garlic from wild onion. Also, in wild onion, the bulbs do not divide to form a new plant as in the case of wild garlic.

Reference
- Wild onion [6,14,24,33,34,36,44,47,67,72,74,76]

Pansy—*Viola tricolor* L.

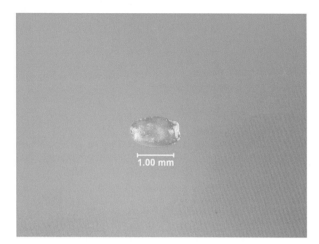

Viola tricolor seed

Field pansy leaves with long petiole

Field pansy flower *Source: Lachlan Cranswick*

Seed
- Size: 1.1–1.3 mm long
- Shape and texture:
- Color: light brown

Field pansy plant

Leaf shape and arrangement—Leaves: Leaf shape varies, with the lower leaves being rounded, and upper leaves narrower at the base than at the tip of the leaves; in a rosette arrangement; leaves are alternate, spoon-shaped, leaf blades taper toward the stalk; young leaves have long *petioles* with rounded teeth on margins; stipules are large "leaf-like," highly divided into segments and are arranged around the stem.

Stolon/Rhizome/Roots: No stolons or rhizomes; stem erect to spreading branched; fibrous root system.

Inflorescence: Flowers are solitary, tiny, irregular in shape, white with pale yellow marking; flowers have three petals and two sepals, the lowest petal having a cup-like projection extending to the back of the flower.

Pansy, field—*Viola tricolor* L.

Other common names: Johnny-jump-up, wild pansy, European field pansy, hearts-ease
Family: Violaceae
Life cycle: Winter annual
Native to: Europe

Distribution and Adaptation
- Native to Europe, in the United States found in most eastern states except extreme north.
- Found in fencerows, gardens, along roadsides, fields, and stream banks.

Morphology/Growth Pattern
- Low growing with an erect growth habit and delicate appearance; often grows in groups; grows up to 12 in. (30 cm) in height; the flat appearance of the flowers and the ring of large, lobed stipules are characteristic for the field pansy.
- Reproduces by seed.

Use and Potential Problems
- Problem weed in turf, pastures, and vegetable and fruit gardens, especially in strawberry fields.

Toxicity
- None recorded.

Similar Species
- Similar species include field *violet* (*Viola arvensis* L.); however, field violet has hairs on the stipules and stems as well as on leaves along the veins on the undersides of the leaves, whereas field pansy lacks hairs; also, field pansy has flower petals approximately three times as long as the sepals.

Reference
- Field pansy [14,24,46−48,58,72,73]

Pennycress, Field—*Thlaspi arvense* L.

Field pennycress seeds

Field pennycress leaves and seedhead *Source: Lachlan Cranswick*

Field pennycress seedhead *Source: Lachlan Cranswick*

Seed
- Size: 1.6−2.0 mm long, 1.1−1.4 mm wide
- Shape and texture: ovate, compressed, appear folded, one long edge has a more pronounced curve than the other; surface covered with a number of roughened lines and resembles a fingerprint impression; narrow groove runs from the hilum to about the middle of the seed
- Color: reddish-brown to black

Field pennycress plant *Source: Lachlan Cranswick*

Leaf shape and arrangement—Leaves: Alternate, simple, smooth, the lower leaves are long stalked (petioled) and have entire margins, while the leaves on the stems are sessile (not petioled) at the arrow-shaped base and have entire with acute to rounded tips and coarsely toothed leaf margins, the leaves on the stem clasp the stem with ear-like projections.

Stolon/Rhizome/Roots: No stolons or rhizomes; stem erect, solitary or several, branching above, glabrous; shallowly taprooted with secondary fibrous root systems.

Inflorescence: Flowers have four white petals and greenish-white sepals on elongated racemes, flower peduncles curved to hold seedpods erect, fruit is flattened, round, and winged on the upper half, and deeply notched at the apex; fruit/pods appear winged and held almost erect or parallel to the stem; fruit produces 4−15 seeds per pod.

Pennycress, field—*Thlaspi arvense* L.

Other common names: Stinkweed, fan-weed, penny-cress, French-weed, bastard cress
Family: Brassicaceae
Life Cycle: Cool-season, winter annual
Native to: Europe

Distribution and Adaptation
- Found throughout the United States and Canada, most common in the northwestern states.
- Found on or near the edges of cultivated fields, in grazed pastures, waste places, and along roadsides.
- Well adapted to nutrient rich soils; adapted to both dry and wet habitats.

Morphology/Growth Pattern
- The seedling develops as a compact vegetative rosette. Stems are erect up to 2—3 ft (0.60—0.90 m) tall, simple or branched above.
- Also known as stinkweed, as the name implies, produces a strong turnip-like odor.
- Reproduces by seed.

Use and Potential Problems
- Troublesome weed in grain fields, overgrazed pastures, nurseries, and in horticultural crops.
- Ill-scented with distinct flavor and odor resembling mustard. When plants or seeds of this plant are eaten by animals, the meat and dairy products give an objectionable, garlicky flavor.

Toxicity
- Glucosinolate, allylthiocyanate.

Similar Species
- Field pennycress is often confused with *shepherd's-purse* (*Capsella bursa-pastoris* L. Shepherdspurse), however, is distinguishable by its triangular seedpods.

Reference
- Field pennycress [12,19,24,26,33,34,39,47,48,61,65,66,72,76]

Pepperweed, Field—*Lepidium campestre* (L.) R. Br.

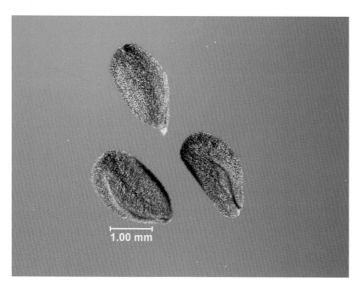

Field peppergrass seeds

Seed

- Size: 2.3–2.5 mm long, 1.1–1.5 mm wide; seedpod is about 6.5 mm long
- Shape and texture: oval-shaped, with one straight side and one strongly convex side, distinct groove that extends most of the length of the seed, translucent tissue on the edges; surface very finely wrinkled, dull
- Color: dark brown to reddish-brown

Field peppergrass seedhead

Leaf shape and arrangement—Leaves: The basal rosette leaves are narrowly obovate and occasionally pinnated toward the base; leaves are opposite, upper leaves/stem leaves are sessile (no petiole) and are clasping at the base on the stem; the leaf margins may be lobed, toothed, or entire; the basal as well as the stem leaves are covered with fine hairs; leaves and stems are hairy.

Stolon/Rhizome/Roots: No stolons or rhizomes; stems are grayish green and covered with fine hairs; stems are simple or branching; the root systems consist of taproots.

Inflorescence: One or more flowering stems that are somewhat erect, numerous flowers on hairy, slender, spreading pedicels/stalks with four small white petals and four green or purplish green sepals that are lanceolate-oblong; flowers bloom toward the apex of the raceme; each flower is about 0.35 cm across; the seedpods are oval, rather boat-shaped, slightly winged and notched at the apex.

Pepperweed, field—*Lepidium campestre* (L.) R. Br.

Other common names: Cream-anther, field pepperwort, field cress, field peppergrass, downy peppergrass
Family: Brassicaceae
Life cycle: Winter annual
Native to: Europe

Distribution and Adaptation
- Found throughout the United States.
- Habitats include cropland, fallow fields, pastures, areas along roadsides and railroads, grassy banks, and waste places.
- The plant grows on various soil types.

Morphology/Growth Pattern
- A winter annual that overwinters as a rosette and produces a flower stem in the spring.
- Grows to about 1−2 ft (0.30−0.60 m) tall, branching at the base and near the apex where the flowers occur.
- Reproduces by seed.

Use and Potential Problems
- No commercial feed values; considered a weed in pasture, hayland, and field crops.

Toxicity
- Glucosinolates.

Similar Species
- Field peppergrass can be distinguished from other *Lepidium* spp. by its hairy stems and leaves. Other members of this genus, like *Virginia pepperweed* (*Lepidium virginicum* L.), and small flowered peppergrass (*Lepidium densiflorum* L.), have stems and leaves that are smooth (glabrous), or nearly so, in addition, Virginia pepperweed does not have clasping stem leaves, unlike those of field pepperweed and the seedpods of Virginia pepperweed are flat and round, unlike the oval and more robust fruit of field pepperweed. Another species, *hoary cress* (*Cardaria draba* L.) is similar in appearance to field peppergrass, however, hoary cress has larger flowers, seedpods without notches, hairless pedicels, and broader leaves. *Field pennycress* (*Thlaspi arvense* L.) is one other member of the mustard family often confused with field peppergrass, similarities include small white flowers in raceme, however, field pennycress has hairless leaves and seedpods that are more flattened than those of field peppergrass, also field pennycress has larger seedpods.

Reference
- Field pepperweed [12,17,19,24,33,47,58,61,66,72,76]

Pepperweed, Virginia—*Lepidium virginicum* L.

Virginia pepperweed seeds

Seed
- Size: 1.6–1.9 mm long, 1.1–1.2 mm wide
- Shape and texture: obovate, tear-shaped, with nearly straight long edge and distinct groove that extends from the small end to the center of each face, radicle tip pointed, thin marginal wing around the radicle edge; surface finely roughened, slight sheen, at high magnification the surface of the seed is uniformly tuberculate
- Color: orange to reddish-yellow

Virginia pepperweed plant

Virginia pepperweed leaves on stem

Pepperweed seedhead *Source: Peter Sforza*

Leaf shape and arrangement—Leaves: Basal leaves of Virginia pepperweed initially forms a rosette, alternate on the stem, are deeply lobed with several small side lobes; leaves on stem are more lanceolate or linear than the basal leaves; 2–10 cm long and 0.5–2 cm wide and leaf size becomes smaller toward the top of the plant; glabrous (no hair), there are no basal leaves after flowers are produced.

Stolon/Rhizome/Roots: No stolons or rhizomes, stems erect, single to much branched, smooth (glabrous) or finely haired; taprooted.

Inflorescence: Small flowers are produced at the end of the stems in a dense cluster that resembles "bottle-brush" or finger-like appearance, individual flowers are inconspicuous and 2 mm wide, four greenish-white petals, two stamens, arranged in elongated clusters; seedpod nearly circular to egg-shaped, flat, notched at tip contains two seeds per pod.

Pepperweed, Virginia—*Lepidium virginicum* L.

Other common names: Poorman's Pepperwort, Poorman's Pepper, Peppergrass
Family: Brassicaceae
Life cycle: Winter or summer annual
Native to: Europe

Distribution and Adaptation
- Virginia pepperweed can be found throughout the United States except Arizona and New Mexico.
- Found in waste places and roadsides as well as in new forests, fields, and disturbed sites.

Morphology/Growth Pattern
- Can grow to 6−24 in. (15−60 cm); the plant initially develops a basal rosette, later produces flowering stems with no basal leaves; the upper leaves are without petioles and the leaf margins may be toothed or entire, plant grows erect and is freely branching.
- Reproduces by seed.

Use and Potential Problems
- No commercial feed values. Virginia pepperweed is a weed of agronomic crops, pastures, vegetables, orchards, and nursery crops.

Toxicity
- Glucosinolates.

Similar Species
- *Field pepperweed* (*Lepidium campestre* (L.) R. Br.) resembles Virginia pepperweed in growth habit; however, field pepperweed leaves clasp the flowering stem and have fruits larger and more robust than those of Virginia pepperweed.

Reference
- Virginia pepperweed [12,17,19,24,44,46,48,58,66,67,72,76]

Pigweed, Redroot—*Amaranthus retroflexus* L.

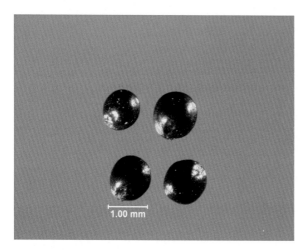

Redroot pigweed seeds

Seed
- Size: 1.0—1.5 mm long, 0.6—1.0 mm wide
- Shape and texture: ovate, lens-shaped, margin slightly notched at hilum from which groove runs inward for short distance; surface shiny, smooth, highly polished (oily)
- Color: black
- Seeds per pound: 600,000

Redroot pigweed plant *Source: James Altland*

Redroot pigweed stem and stem *Source: James Altland*

Leaf shape and arrangement—Leaves: Egg-shaped, wavy-margined, alternate leaves, leaf tips slightly notched, leaves hairy and have prominent veins especially on the lower surfaces; bristly bracts.

Stolon/Rhizome/Roots: No stolons or rhizomes; look for red tinge on lower stem.

Inflorescence: Flowers are numerous, small, greenish with stiff, sharp-pointed bracts under them; densely crowded in large, bristly, simple, or branched, terminal, or axillary clusters.

Redroot pigweed leaves, stem, and seedhead

Pigweed, redroot—*Amaranthus retroflexus* L.

Other common names: Green amaranth, rough pigweed, amaranth pigweed, careless weed, red-root, wild beet
Pigweed, smooth—*Amaranthus powellii* S. Wats.
Family: Amaranthaceae
Life cycle: Warm-season, summer annual
Native to: Tropical America

Distribution and Adaptation
- Widely distributed through the United States and southern Canada.
- Thrives in warm climates.
- Found in cultivated fields, pastures, roadsides, and waste places.

Morphology/Growth Pattern
- The stem is light green, erect, stout, tough, rough-hairy, much branched and 1−6 ft (0.30−1.86 m) tall with a long, somewhat fleshy, red taproot (red root pigweed).
- A prolific seed producer, pigweed will produce up to 100,000 seeds per plant.
- Seeds stay viable in the soil for many years.
- Reproduces by seed.

Use and Potential Problems
- Common weed in pastures, row crops, turf, and ornamentals.
- Birds eat pigweed seeds and deer browse the foliage.
- Seeds can be easily removed from crop seeds due to their small size.
- Poisonous to livestock, also acts as a host for the beet leafhopper.

Toxicity
- None recorded—believed to be present in leaves and seeds.

Similar Species
- *Smooth pigweed* (*Amaranthus powellii* S. Wats.) is very similar to redroot pigweed; however, this species has terminal panicles that appear less dense, compact, and bristly than those of redroot pigweed. Additionally, the bracts of smooth pigweed are only slightly longer than the sepals, unlike those of redroot pigweed.

Reference
- Redroot pigweed [12,17,19,24,33,44,48,58,61,65−67,72,76]

Pineappleweed—*Matricaria discoidea* DC.

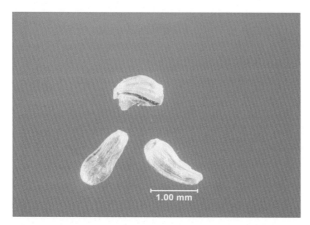

Pineappleweed seeds

Pineappleweed leaves and stem

Pineappleweed flowers *Source: James Altland*

Seed
- Size: 1.3—1.6 mm long, 0.5—0.7 mm wide
- Shape and texture: oblong, four to five angled, elongated, and often curved, narrow toward the base; surface striate, slightly sheen
- Color: light brown with reddish brown line on sides

Pineappleweed plant

Leaf shape and arrangement—Leaves: When young forms a rosette; stem leaves opposite, simple, linear (but wider at the base), fern-like and sessile or with very short *petiole*, leaves very divided, up to 6 cm long and to 0.5 cm wide; leaf tip tapers to points; the base of each pair of leaves wraps around the stem in a sheath; leaves mostly glabrous but with some patches of fine white hairs beneath each opposite pair of leaves.

Stolon/Rhizome/Roots: No stolons or rhizomes; stem ascending to erect, slender, round, smooth, glabrous or with a few sparse hairs, branched, to 15 + cm tall; taprooted.

Inflorescence: Flowers numerous, greenish-yellow disk (tubulate) florets; solitary on short flower stalk (peduncles) on the top of the stems, cone-shaped heads; pineappleweed has no ray flowers, each flower head is surrounded by several overlapping bracts with papery edges; fruits are yellowish gray, each 1—1.5 mm long and have a small tubercle in the middle.

Mayweed Chamomile—*Anthemis cotula* L.

Mayweed chamomile seeds

Mayweed chamomile flower *Source: James Altland*

Seed
- Size: 1.3−1.8 mm long, 0.7−1.0 mm wide
- Shape and texture: wedge-shaped, with several roughened longitudinal ribs
- Color: light brown

Mayweed chamomile plant *Source: James Altland*

Leaf shape and arrangement—Leaves: The leaves in the first leaf pair are opposite and can be nearly hairless, the subsequent leaves, however, are alternate, are lobed to deeply dissected, nearly hairless or hairy; approximately 19−60 mm long and 25 mm wide.

Stolon/Rhizome/Roots: No stolons or rhizomes; stem ascending to erect, slender, round, smooth, glabrous or with a few sparse hairs, branched, to 15 + cm tall; taprooted.

Inflorescence: Daisy-like flower head with 12−15 white, ray flowers, the rays are three-toothed and are close to the stalk, in the center yellow disc flowers, flower 19 mm across.

Pineappleweed—*Matricaria discoidea* DC.

Other common names: Disc mayweed, pineapple mayweed, rayless chamomile
Family: Caryophyllaceae

Mayweed chamomile—*Anthemis cotula* L.

Other common names: Pineappleweed, dog fennel, mayweed, stinking chamomile, mayweed
Family: Asteraceae
Life cycle: Summer or winter annual
Native to: Europe

Distribution and Adaptation
- Distributed throughout the United States.
- Found in pastures, field crops, waste places, abandoned fields, along roadsides and railroads, and woodlands as well as infrequently mowed grasses.
- This plant tolerates compacted soils and frequent mowing.
- It thrives in low fertility soils or areas where other plants cannot grow.

Morphology/Growth Pattern
- The name pineappleweed derived from its "pineapple-like" aroma, which is produced when the plant is crushed or injured and the shape of the flowers resembles that of cultivated pineapple.
- The key morphological characteristic of pineappleweed is the rosettes that are formed when the plant is at its vegetative stage; leaves are highly subdivided into narrow, feathery segments, pubescent to nearly glabrous and succulent; the stem is erect, stiff, and forks into branches; nodes swollen; cone-shaped or rounded small, yellow-greenish flowers with no ray flowers.
- Mature pineapple-weed can grow up to 3—12 in. (7.5—30 cm) tall.

Use and Potential Problems
- Primarily a weed of landscapes, nurseries, and turfgrass, but also occurs in compacted areas like gravel roads or walkways.
- Due to its less competitive nature, the species survival with many perennial broad-leaved plants will decline overtime.
- Reproduces by seed.

Toxicity
- None recorded.

Similar Species
- *Mayweed chamomile* (*Anthemis cotula* L.) looks similar to pineappleweed, however, mayweed chamomile has an offensive odor and white ray flowers and is relatively taller than pineappleweed; *dogfennel* (*Eupatorium capillifolium* L.) is the other plant confused with pineappleweed; however, like mayweed chamomile, this plant also does not produce pineapple-like odor and is much taller than pineappleweed.

Reference
- Pineappleweed/Mayweed [17,24,33,48,58,61,65—67,72,76]

Pink, Deptford—*Dianthus armeria* L.

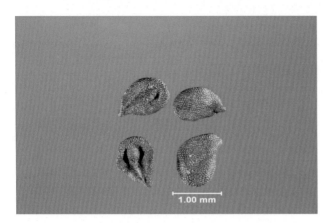

Deptford pink seeds

Seed

- Size: 1–1.5 mm long
- Shape and texture: kidney-shaped, compressed and wrinkled with small bumps across the surface
- Color: dark gray

Deptford pink plant *Source: Wise Co. VA*

Deptford pink hairy stem and leaves *Source: Wise Co. VA*

Leaf shape and arrangement—Leaves: Opposite, entire; numerous leaves at the base of the stem, further up on the stem, however, the paired narrow, long, pubescent stem leaves are arranged far apart on the stem; lower leaves are larger and slightly broader; leaves can reach 5 cm in length.

Stolon/Rhizome/Roots: No stolons or rhizomes; stems simple or clustered, pubescent; taprooted.

Inflorescence: The flowers are clustered at the top of the stem and are surrounded by long, stiff, leafy flower bracts; the flower petals are deep pink dotted with white spots; flowers have five narrow regular parts and are up to 1.3 cm wide.

Deptford pink stem and flower *Source: Wise Co. VA*

Pink, deptford—*Dianthus armeria* L.

Other common names: Grass pink
Family: Caryophyllaceae
Life cycle: Summer annual or biannual
Native to: Europe

Distribution and Adaptation
- Found throughout the United States.
- Found in grasslands, fencerows, gardens, lawns, and along roadsides.

Morphology/Growth Pattern
- Nonnative, annual, herbaceous upright, grass-like plant which can reach 20 in. (50 cm) in height; the leaves are rosette and are green which distinguish this plant from other members of its family which have gray-green leaves; the flower petals are pinkish with dots on them.
- Reproduces by seed.

Use and Potential Problems
- No commercial feed value; considered a weed.

Toxicity
- None recorded.

Similar species
- Deptford pink can be easily identified in the field because of the distinct appearance of the flower petals; the petals of deptford pink are usually narrower than the petals of other *Dianthus* spp., their outer edges are toothed, and they have small white dots across the surface. The flowers of this species are smaller in size and less showy than the flowers of *Dianthus* spp. (Pinks) that are commonly cultivated in flower gardens. The common name refers to an area of England where this species was once common.

Reference
- Deptford pink [17,33,47,48,61,66,76]

Plantain, Blackseed—*Plantago rugelii* Decne.

Blackseed plantain seeds

Seed

- Size: 2.0−2.6 mm long, 0.9−1.1 mm wide
- Shape and texture: highly irregular, frequently elliptical or oval with angular margins, dorsal side convex but not as much so as the ventral side, white seed scar (hilum) near center on one side; surface very finely wrinkled, dull
- Color: black or dark brown with a broad, lighter colored, longitudinal band down the middle of the dorsal side

Blackseed plantain plant *Source: NCSU weed*

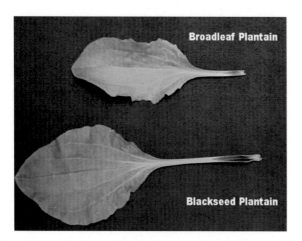

Broadleaf plantain (top) and blackseed plantain (bottom) *Source: NCSU weeds*

Blakseed plantain seedhead *Source: NCSU weed*

Leaf shape and arrangement—Leaves: All three plantains have "basal rosette" leaves with parallel venations running lengthwise; blackseed plantain has broad, oval, rounder leaves with wavy toothed margins, usually hairless and has thick stalk with purplish petiole bases.

Stolon/Rhizome/Roots: No stolons or rhizomes; fibrous roots are produced from a thick, tough, short taproot-like woody underground stem; leafless stems.

Inflorescence: Flowers are produced on unbranched stalks rising from the rosette; flowering stems are 13−38 cm long, clustered with small flowers that have whitish petals and bracts surrounding the flowers.

Plantain, Buckhorn—*Plantago lanceolata* L.

Buckhorn plantain seeds

Seed

- Size: 2.3−2.6 mm long, 1.0−1.3 mm wide
- Shape and texture: long oval shape (boat-shaped), compressed, slightly concave−convex with the margins rolled toward the concave side, hilum is placed in the center of the concave face; surface is smooth without magnification, under magnification appears finely roughened
- Color: seeds are light to dark brown with a broad, yellowish, longitudinal stripe in the middle of the dorsal side; dark hilum

Buckhorn plantain *Source: NCSU*

Buckhorn leaves *Source: NCSU*

Leaf shape and arrangement—Leaves: All three plantains have "basal rosette" leaves with parallel venations running lengthwise. Buckhorn plantain leaves are linear to lanceolate, slender (grass-like in basal), often twisted or curled, with smooth margins, may be sparsely hairy or without hairs.

Stolon/Rhizome/Roots: No stolons or rhizomes; fibrous roots are produced from a thick, tough, short taproot-like woody underground stem; leafless stems.

Inflorescence: The seedhead is dense, tapered, white to tannish flower spike-like stalk, 4−25 in. long, flowers have four sepals and four petals, at maturity spike-like pins in *pin-cushion*.

Buckhorn plantain seedheads *Source: NCSU*

Plantain, Bracted—*Plantago aristata* Michx

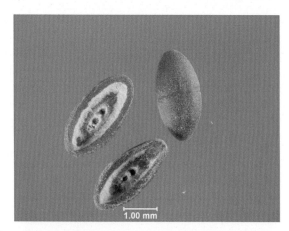

Bracted plantain seeds

Bracted plantain, buckhorn plantain, and blackseed plantain (left to right)

Bracted plantain seedhead *Source: NCSU weed*

Seed

- Size: 2.5−3.0 mm long, 1.3−1.5 mm wide
- Shape and texture: oval to ovate, boat-shaped; surface smooth, white-margined seed scars in the center with alternating elliptical bands of caramel brown in color
- Color: brown and white

Bracted plantain plant *Source: NCSU*

Leaf shape and arrangement—Leaves: All three plantains have "basal rosette" leaves with parallel venations running lengthwise. Bracted plantain has long, slender, upright, grass-like, linear to very narrowly oblanceolate leaves with long tapering tips, dark green, hairy or hairless.

Stolon/Rhizome/Roots: No stolons or rhizomes; fibrous roots are produced from a thick, tough, short taproot-like woody underground stem; leafless stems.

Inflorescence: Flower spikes have conspicuous hairy leaf-like bracts coming out of the seedhead, the bracts are two to eight times as long as the flower.

Plantain, blackseed—*Plantago rugelii* Decne.

Other common names: Rugel's plantain, broadleaved plantain, pale plantain, silk plantain, white man's foot, purple-stemmed plantain
Family: Plantaginaceae

Plantain, buckhorn—*Plantago lanceolata* L.

Other common names: English plantain, narrow-leaved plantain, rib-grass, ribwort, buck plantain, black-jacks
Life cycle: Cool-season perennial

Plantain, bracted—*Plantago aristata* Michx.

Other common names: Rat-tail plantain, western buckhorn, western ripplegrass
Life cycle: Cool-season, winter annual
Native to: North America

Distribution and Adaptation
- Found throughout the United States, Canada, and Mexico.
- Commonly found in drier sites and neutral to basic soils; persists under close cutting or grazing (blackseed and buckhorn plantains). Bracted plantain is commonly found on sandy, droughty soils.

Morphology/Growth Pattern
- All the plantains are characterized by parallel venations that run lengthwise, however, bracted plantain lacks distinct parallel venations.
- Flowering stalks of blackseed plantain, buckhorn plantain, and bracted plantain can grow up to the height of 10 in. (25 cm), 2.5 ft (0.80 m), and 6−12 in. (15−30 cm), respectively.
- Reproduces by seed.

Use and Potential Problems
- No commercial feed value. Primarily weed of turfgrass, landscapes, orchards, nursery crops, and pastures.
- Reproduces by seed.

Toxicity
- None recorded.

Similar Species
- Blackseed plantain is often confused with *broadleaf plantain* (*Plantago major* L.). Blackseed plantain is similar to broadleaf plantain except blackseed plantain stems and petioles are longer and the leaves somewhat larger; the leaves of blackseed plantain are hairless, and have toothed and wavy margins, unlike the leaves of broadleaf plantain. Additionally, blackseed plantain leaves tend to be light green, less waxy, more tapered at the tip, and more red to purple at the base of the petiole.

Reference
- Plantains [14,17,18,24,33,44,46,48,58,61,65−67,71,76]

Pokeweed—*Phytolacca americana* L.

Pokeweed seeds

Pokeweed leaves (Pokeweed red stem and leaf)

Pokeweed flower *Source: Peter Sforza*

Pokeweed seedhead

Seed
- Size: 2.8−2.9 mm long, 2.5−2.8 mm wide
- Shape and texture: nearly round and flattened with a small extension and a V-shaped notch, may be depression in the center of each face, hilum a broad notch partially filled with corky tissue; surface very smooth very glossy
- Color: black with purplish undertones

Pokeweed plant

Leaf shape and arrangement—Leaves: Egg-shaped, alternate, entire, smooth, *petiolated*, and taper at the end; leaves get smaller in size toward the top of the plant; without hairs; mature leaves give off a pleasant odor when crushed.

Stolon/Rhizome/Roots: No stolons or rhizomes; stems erect, large, smooth, are often purple-tinged and branch from the root crown at the base of the plant; large white tap root up to 15 cm in diameter.

Inflorescence: The white flowers are borne in 15−25 cm racemes (nodding cluster); flowers have five white to pink-tinged sepals; fruits are green when young and turn dark-purple to black when mature, berries contain dark red juice.

Pokeweed—*Phytolacca americana* L.

Other common names: Pokeberry, Virginia poke, poke, inkberry, pigeon berry, garget, red ink plant, American cancer, cancer jalap, poke salad
Family: Phytolaccaceae
Life cycle: Cool-season perennial
Native to: United States

Distribution and Adaptation
- A native plant throughout eastern North America from Ontario, Quebec, and Maine, west to Minnesota, and south to Texas, Mexico, and Florida.
- Grows in waste places, along roads, fencerows, in abandoned fields, forest or natural areas, in disturbed areas, fields, and clearings.
- Grows on deep, rich, as well as gravelly, and sandy soils.

Morphology/Growth Pattern
- The plant can reach 3—10 ft (0.90—3 m) tall, succulent, coarse, tree-like with egg-shaped leaves, purple-tinged stems, and dark purple berries.
- The fruits resemble the berries of nightshade and thus pokeweed is sometimes called American nightshade.
- Where adapted, pokeweed self seeds readily can be propagated by taproot.

Use and Potential Problems
- Can be a troublesome weed but can be controlled easily.
- Some gardeners use pokeweed in shrub or mixed borders.
- All parts of the plant (roots, berries, seeds, and mature stems and leaves) are dangerously poisonous to cattle, horses, swine, and humans, especially the roots. Swine are most often affected since they are capable of "grubbing-up" the roots.

Toxicity
- The toxic elements in pokeweed are phytolaccatoxin and related triterpene saponins, an alkaloid (phytolaccin), and histamines.
- Symptoms include burning of mouth and throat, salivation, severe stomach irritation, vomiting, bloody diarrhea, spasms, and convulsions; can be fatal.

Reference
- Pokeweed [6,12,24,33,46,58,62,67,72]

Prickly, Sida—*Sida spinosa* L.

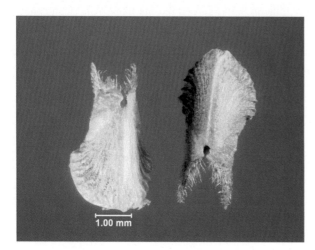

Prickly sida seeds

Seed
- Size: 1.8−2.2 mm long (excluding spines), 1.3−1.7 mm wide
- Shape and texture: three-angled, has spine-like projections at the apex, projections are granular and their inner margin bears short spines; surface covered with a network of veins and with large shallow spaces between veins
- Color: dark reddish-brown

Prickly sida plant *Source: Nathan O'Berry*

Prickly sida stem, leaves, and flower head

Leaf shape and arrangement—Leaves: Oval to lanceolate with toothed margins; alternate, simple; have long *petioles* and small spines (stipules) that are 5−8 mm long at the base of each leaf petiole; spines or "prickles" form at the base of the leaf stems.

Stolon/Rhizome/Roots: No stolons or rhizomes; a long slender taproot and a fibrous root system.

Inflorescence: Flowers are at the axils of the leaves and have pale yellow petals, the seedpod splits into *five sections*, each with one seed and topped with two spines (two sharp spreading beaks at the top).

Prickly, sida—*Sida spinosa* L

Other common names: Spiny sida, prickly mallow, thistle mallow, false mallow, Indian mallow, teaweed
Family: Malvaceae
Life cycle: Warm-season, summer annual
Native to: Tropical America

Distribution and Adaptation
- Found throughout the eastern United States, extending north to Massachusetts and Michigan and west to Nebraska; however, most common in the southeastern United States.

Morphology/Growth Pattern
- The branching, upright growth of this plant seldom reaches more than 1 foot (0.30 m); leaves are alternate along the stem, oval to lanceolate with toothed margins along the edges, small spines at the base of each leaf and branch; leaves are approximately 0.75−2 in. (1.9−5 cm) long, inconspicuously hairy; leaves occur on petioles that are 0.5−1.25 in. (1.2−3.2 cm) long and have small spines (stipules) at the base of each leaf petiole; stems erect, branched, and hairy; flowers occur singly or in clusters on flower stalks (peduncles) that arise from the area between the stems and leaf petioles. Flowers consist of five yellow petals that are 4−6 mm long.
- Reproduces by seed.

Use and Potential Problems
- This weed is one of the 10 most common and troublesome weeds in peanuts, cotton, and soybeans in most of the southern states.
- Prickly sida is primarily a weed of agronomic crops, but can also be found in horticultural crops, landscapes, pastures, hay fields, and gardens.

Toxicity
- None recorded.

Similar Species
- *Velvetleaf* (*Abutilon theophrasti* Medik.), *spurred anoda* (*Anoda cristata* L.), and *arrowleaf sida* (*Sida rhombifolia* L.) seedlings are very similar to those of prickly sida. However, prickly and arrowleaf sida have two heart-shaped cotyledons unlike the round and heart-shaped cotyledons of velvetleaf. Spurred anoda also has two heart-shaped cotyledons like prickly and arrowleaf sida, however, the first true leaf of spurred anoda is not as coarsely toothed as that of prickly or arrowleaf sida. The cotyledons of arrowleaf sida are essentially identical to those of prickly sida, however, the first true leaf of arrowleaf sida is diamond-shaped in outline and tapers to the base unlike the first true leaf of prickly sida.

Reference
- Prickly sida [17,19,24,33,58,66,67,76]

Primrose, Evening—*Oenothera biennis* L.

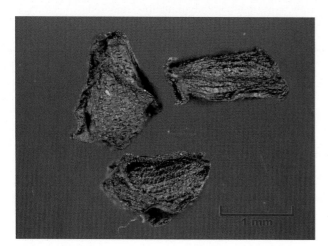

Evening primrose seeds

Seed
- Size: 1.1–1.5 mm long, 0.9–1.1 mm wide
- Shape and texture: oval in outline, dorsal side convex, ventral side appears as convex or two faced with broad, rounded, central ridge; seed surface dull and granular, several fine lengthwise wrinkles on each side of the seed
- Color: reddish-brown with two distinguished brown spots at the bases of one face of the seed

Evening primrose plant *Source: John Wright*

Evening primrose flower *Source: Wise Co., VA*

Leaf shape and arrangement—Leaves: Alternate spiraling ascending or spreading around the stem; closely spaced, lanceolate; entire to lobe, sparsely pubescent; long, wide at the base and taper to a point; sessile or short petioled; sparsely pubescent; 5–15 cm long and 1.5–4 cm wide; first year of growth leaves form rosette on the ground, rosette leaves are long-petioled; the tall flowering stalk is produced the second year.

Stolon/Rhizome/Roots: No stolons or rhizomes; stems are woody, stout, and are covered with soft silky hairs; taprooted.

Inflorescence: The bright yellow flowers become orange in age, flowers on terminal spike several to many, opens in the evening; the four broad petals with notched tips are produced on short-stalk in the axils of the leaves and are 1–2.5.0 cm long; with yellow anthers extended and visible; sepals 1.5–5.0 cm long.

Primrose, evening—*Oenothera biennis* L.

Other common names: Common evening primrose, field primrose, tree primrose, willow herb, night willow herb, fever plant, coffee plant
Family: Primulaceae/Onagraceae
Life cycle: Cool-season, winter annual, or biennial
Native to: United States

Distribution and Adaptation
- Distributed from the East Coast to the eastern edge of the Great Plains.
- Found waste places, fields, meadows, and along roadsides, early invader of new forest plantations.

Morphology/Growth Pattern
- This biennial, herbaceous, erect plant has a fleshy, biennial root which produces a rosette of leaves the first year; leaves are stalked, elliptical in shape with irregular margins; flowering stem up to 6 ft (1.8 m) tall the second year; the leaves on the stem are sessile and have toothed margins; flowers are in long spikes or racemes at the end of stem or branches and in the axils of reduced leaves.
- Reproduces by seed.
- Seed remains viable in the soil for decades.

Use and Potential Problems
- Mainly considered as weed with little economic importance.

Toxicity
- None recorded.

Similar Species
- *Cutleaf evening-primrose* (*Oenothera laciniata* Hill.) and evening primrose are similar in appearance in the rosette stage. Cutleaf evening-primrose stem is hairy, often reddish and prostrate to ascending from the base with an erect tip; the leaves of cutleaf primrose are toothed and are irregularly lobed.

Reference
- Evening primrose [19,36,39,46−48,58,61,66,72,76]

Purslane, Common—*Portulaca oleracea* L.

Common purslane seeds

Common purslane stem and leaves

Common purslane flower *Source: James Altland*

Seed
- Size: 0.6−0.9 mm long, 0.4−0.6 mm wide
- Shape and texture: ovate to rounded triangular, with a small notch on one side just below the small end, each face has a shallow groove extending from the small end of the seed toward the center; surface rough due to small rounded projections in curved rows on the surface of the seed, shiny
- Color: black

Common purslane plant *Source: James Altland*

Leaf shape and arrangement—Leaves: Oval, shiny, opposite, or alternate (near the base) along the stem and are without petioles (sessile); leaves are clustered at the ends of the branches; leaf size varies from 1.2 to 5 cm in length; very succulent (juicy), hairless, often tinged red.

Stolon/Rhizome/Roots: No stolons or rhizomes; stems are hairless, thick, round, fleshy, very succulent and reddish-brown; branch from the central root; taproot with fibrous secondary roots.

Inflorescence: Small, yellow, four to six rounded petals, born singly or in clusters of two or three in stem axils or at tips of stems, or are several together in the leaf clusters at the ends of the branches; seeds are borne in small capsules/pod with many seeds.

Purslane, common—*Portulaca oleracea* L.

Other common names: Pusley, kitchen purslane, garden purslane, wild portulace, fatweed
Family: Portulacaceae
Life cycle: Summer annual
Native to: Asia/or Europe

Distribution and Adaptation
- Abundant throughout the world; in the United States, it is most abundant in the northeastern states, least common in the Pacific Northwest.
- Thrives under dry conditions but also grows well in irrigated fields as well as warm, moist conditions.
- Prefers loose, nutrient-rich, sandy soil and is not shade tolerant.
- Very drought resistant, this is attributed to its succulent nature with a water content of over 90%.

Morphology/Growth Pattern
- The succulent reddish to flesh-colored stems originate from a central rooting point producing watery and sticky sap; the plant branches at the base and along the stems and can form large circular mats; the stems arise from a single taproot.
- The wild plant grows prostrate, often forms a dense mat or grows up to 1 foot (0.30 m) tall while the cultivated forms are more upright and vigorous than the wild type.
- Flowers usually open only in sunshine.
- Reproduces from seed and stem fragments.

Use and Potential Problems
- Weed of cultivated fields and gardens, pastures, barren driveways, waste places, eroded slopes, low-maintenance lawns, ornamental plantings, commercial orchards, and vegetable crops.
- Due to its ability to produce a large number of seeds, common purslane can rapidly colonize any warm, moist site.
- Soil disturbance by tillage increases germination and rerooting of common purslane.

Toxicity
- Oxalates.

Similar Species
- Due to the prostrate growth habit of common purslane, it is often confused with the *spurges* (*Euphorbia* spp.) or with *prostrate knotweed* (*Polygonum aviculare* L.); however, the spurges release a milky sap when injured and prostrate knotweed is not succulent and has a papery sheath called ocreae around the leaf base (node).

Reference
- Common purslane [12,16,19,21,26,33,47,58,61,65−67,72,74,79]

Ragweed, Common—*Ambrosia artemisiifolia* L.

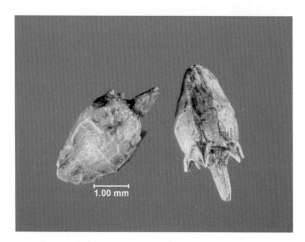

Common ragweed seeds

Common ragweed leaf *Source: NCSU weed*

Common ragweed leaves and flower heads

Seed

- Size: 3.0—4.0 mm long, 1.8—2.5 mm wide
- Shape and texture: both types have long, spiny central protuberance (crown-like) which is surrounded by a circle of five to eight projections that extend downward as ribs; protuberances are slender and sharper in common ragweed seeds, while they are thicker and more blunt in giant ragweed; surface striate, dull, interrib spaces may have ridge, be wrinkled or roughened
- Color: grayish-brown, yellowish-brown, brown or reddish-brown

Common ragweed plant *Source: NCSU weed*

Leaf shape and arrangement—Leaves: Opposite, but upper alternate, leaves are like carrot, deeply lobed or dissected (7—10), smooth or hairy.

Stolon/Rhizome/Roots: No stolons or rhizomes; stems are simple or branching, hairy or smooth; shallow taproot.

Inflorescence: The flowers of both ragweeds are similar but giant ragweed flower heads are larger than common ragweed; male and female flowers are in separate heads (the female heads are small and are in the leaf axils, the more showy male heads are in small, inverted clusters at the top of the plant); with the pollen-producing heads being the more conspicuous in long racemes, greenish and drooping. These produce a large amount of pollen; the seed-producing flowers are one to several in the axils of the upper leaves and are inconspicuous when young; flowers are yellow.

Ragweed, Giant—*Ambrosia trifida* L.

Giant ragweed seed

Giant ragweed stem *Source: Wise Co., VA*

Giant ragweed stem *Source: Wise Co., VA*

Seed
- Size: 5.6–6.3 mm long, 3.0–3.8 mm wide shape and texture: obovate with stout spreading spines and two incurved teeth at the apex, both types have long, spiny central protuberance (crown-like) which is surrounded by a circle of five to eight projections that extend downward as ribs, interrib spaces may have ridge; surface lower portion of the seed is smooth, papery, wrinkled, or roughened
- Color: grayish-brown, yellowish-brown, brown or reddish-brown

Giant ragweed plant

Leaf shape and arrangement—Leaves: Opposite, long-stalked, leaves distributed in pairs at nodes and usually have three lobes, deep clefts except in the inflorescence, and distinctive leaf serrations.

Stolon/Rhizome/Roots: No stolons or rhizomes; has rough, hairy stems; deep tap root.

Inflorescence: The flowers of both ragweeds are similar but giant ragweed flower heads are larger than common ragweed (see common rag weed inflorescence description).

Ragweed, common—*Ambrosia artemisiifolia* L.

Other common names: Roman wormweed, hogweed, annual ragweed, small ragweed, wild tansy, hayfever, bitterweed, hayweed, carrotweed, annual ragweed
Family: Asteraceae/Compositae

Ragweed, giant—*Ambrosia trifida* L.

Other common names: Kinghead, horse cane, horseweed, greater ragweed, great ragweed, big ragweed, buffaloweed, tall ambrosia, crownweed, wild hemp, big bitterweed, tall ragweed
Life cycle: Warm-season, summer annuals
Native to: North America

Distribution and Adaptation
- Common ragweed is distributed in the northwestern and southeastern United States, also in Canada, Central and South America, the West Indies, South Pacific, and Australia; giant ragweed is also found in the same area but is less common than common ragweed.

Morphology/Growth Pattern
- Common ragweed has leaves that are deeply divided and covered with short white hairs; stems are stout, coarse, simple, or branching, smooth with long hairs; can reach up to 1−6.6 ft (0.30−2 m) in height; giant ragweed grows up to 7−20 ft (2−6 m) in height; flowers are small and produce yellow pollen; seeds are developed in bell-shaped clusters.
- Viable seeds can persist in the soil for many years.
- Reproduces by seed.

Use and Potential Problems
- Considered a weed, however, is known to have high nutritive forage value; early inhabitant of disturbed soils.
- All the ragweeds produce abundant amount of pollen that is the most frequent cause of hay-fever in late summer/ fall.
- Commonly found in pastures, crop fields, roadsides, and waste places.

Toxicity
- None recorded.

Similar Species
- *Common ragweed* (*Ambrosia artemisiifolia* L.) is easily distinguished from giant ragweed by its leaves that are divided into deep and narrow segments (leaves somewhat resemble those of *wild carrot* (*Daucus carota*)). Giant ragweed leaves are unlobed to deeply lobed to three to five sections.

Reference
- Ragweeds [14,17,19,24,33,34,39,44,46−48,57,58,61,65−67,72,74,76]

Scarlet Pimpernel—*Anagallis arvensis* L.

Scarlet pimpernel leaves and flowers

Scarlet pimpernel plant

Scarlet pimpernel flower

Leaf shape and arrangement—Leaves: Arranged opposite from each other, may appear in whorls of three, oval to elliptic in outline, 2.5 cm in length without the petiole; lower surface of leaves have small dark purple spots; leaves with or without hairs.

Stolon/Rhizome/Roots: No stolons or rhizomes; stem square branching from the base reaching up to 25 cm.

Inflorescence: Long stalk, solitary arise from the leaf axils, on long peduncle, nodding flower; flower have five petals with tiny hairs along the margins, flower color ranging red-orange to occasionally white or blue; flower 0.6 cm across.

Scarlet pimpernel—*Anagallis arvensis* L.

Family: Primulaceae
Life cycle: Summer annual
Native to: Europe

Distribution and Adaptation
- Distributed throughout the United States.
- Found in waste places and roadsides; prefers sandy type soil.

Morphology/Growth Pattern
- Low growing annual, height ranging from 4 to 12 in. (10−30 cm).
- Flowers open only under full sun sunlight, closes late afternoon and cloudy weather conditions.
- Spreads by seeds.

Use and Potential Problems
- Mostly weed of turfgrass and landscapes.

Toxicity
- Leaves may cause severe dermatitis.

Similar Species
- Scarlet pimpernel easily confused with *common chickweed* (*Stellaria media* L.); however, scarlet pimpernel has square stems and reddish-orange flowers, while common chickweed has round stems and white flowers. Scarlet pimpernel can also be easily identified by the purple spots on the lower surface of the leaves.

Reference
- Scarlet pimpernel [12,24,48,71,75]

Shepherd's-Purse—*Capsella bursa-pastoris* (L.) Medik.

Shepherd's-purse seeds

Shepherd's-purse flowers and seedhead *Source James Altland*

Shepherd's-purse seedpods *Source: James Altland*

Seed
- Size: 0.9—1.2 mm long, 0.4—0.6 mm wide; smaller than greenflower pepperweed
- Shape and texture: oblong, flattened, two longitudinal grooves, straight crease; surface smooth, dull, under high magnification it appears very finely textured and has a slight sheen
- Color: dull orange, the basal end of the seed is darker

Shepherd's-purse Plant *Source: James Altland*

Leaf shape and arrangement—Leaves: Basal rosette leaves are on a stalk and are rounded, the shape and the size of the rosette leaves are variable, the older leaves are deeply toothed; the leaves on the flower stalk are small, slightly toothed, alternate and clasp the stem; upper leaves are spear-shaped with pointed auricles; rosette has deeply serrated lower leaves, with sparse hairs on surface, leaf tips point up.

Stolon/Rhizome/Roots: No stolons or rhizomes; stem erect; taproot (may be branched) with secondary fibrous root system.

Inflorescence: The small white flowers appear in clusters at the top of the stalk, petals and sepals are arranged in fours, similar to all the mustard flowers, they have four petals which form a cross and six stamens; the seedpod is heart-shaped (purse-shaped) or triangular and each fruit has two cells containing many small seeds, the triangular seedpod is approximately 4—8 mm long and notched at top; peduncle longer than greenflower pepperweed.

Shepherd's-purse—*Capsella bursa-pastoris* (L.) Medik.

Other common names: Shepherd's purse, shepherd's-bag, shepherd's-scrip, shepherd's-sprout, lady's purse, witches pouches, rattle pouches, case-weed, pick-pocket, pick-purse, blind-weed, pepper-and-salt
Family: Brassicaceae
Life cycle: Cool-season, winter annual
Native to: Europe

Distribution and Adaptation
- Found all over the world except in tropical climates.
- Best adapted and widely found in sunny, moist to dry, rich, disturbed soil, but it will also grow in partially shaded, extremely poor soils.
- The seeds live in the soil for a long period of time.

Morphology/Growth Pattern
- Shepherd's-purse grows from a basal rosette, with erect stems that grow 3—18 in. (8—46 cm) tall.
- The plants die soon after fruiting.
- One of the first broadleaf weeds appearing in the spring.
- Has a distinctive and not quite pleasant odor.
- Reproduces by seed.

Use and Potential Problems
- A weed of cultivated fields (wheat fields), pastures especially alfalfa, turf, gardens, cultivated vegetables, also found along roadsides and waste areas.
- Although not a very competitive weed, if the crop stand is weak, it can quickly occupy an area and could cause a reduction in crop yield and quality.
- A variety of birds, including grouse and goldfinches, eat the seeds; chicken and dairy cattle also consume the plant; however, the flavor of eggs and milk may be affected due to the strong flavor of shepherd's-purse.

Toxicity
- Oxalates.

Similar Species
- The rosette basal leaves of shepherd's-purse resemble the basal leaves of *dandelion* (*Taraxacum officinale* F.H. Wigg aggr. G.H. Weber, ex Wigger.); however, the leaf tips of dandelion point down, while those of shepherd's-purse points up. There are also similarities in appearance between shepherd's-purse and Virginia pepperweed, however, the fruits of Virginia pepperweed are flat and round not triangular to heart-shaped like shepherd's-purse.

Reference
- Shepherd's-purse [12,14,17—19,24,33,46,47,58,61,66,67,72,76]

Smartweed, Pennsylvania—*Persicaria pensylvanica* (L.) M. Gómez; Ladysthumb or Smartweed—*Polygonum persicaria* L. = *Persicaria maculosa* Gray

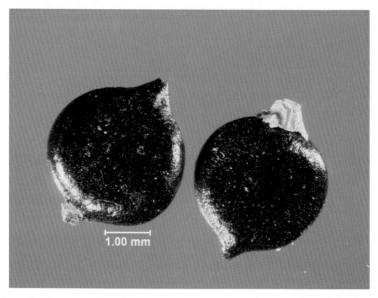

Pennsylvania smartweed seeds

Seed

- Size: 2.8–3.5 mm long, 2.6–3.0 mm wide
- Shape and texture: nearly round, flattened with short, pointed tip; shaped as "ace of spades"; surface smooth, highly glossy, under high magnification it appears very finely wrinkled
- Color: black to dark-brown with very pale attachment point

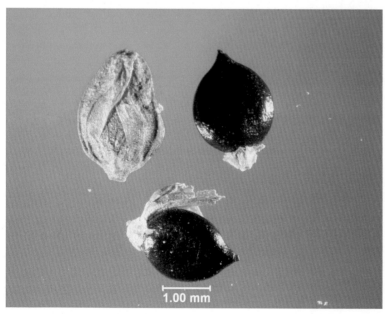

Ladysthumb seeds

Seed

- Size: 2.0–3.2 mm long, 1.8.–1.9 mm wide, smaller than Pennsylvania smartweed
- Shape and texture: short, ovate with a small apical tip, one face may have basal swelling; surface smooth, highly glossy
- Color: black to dark-brown with very pale attachment point

Smartweed, Pennsylvania—*Persicaria pensylvanica* (L.) M. Gómez
Ladysthumb or Smartweed—*Polygonum persicaria* L. = *Persicaria maculosa* Gray

Pennsylvania smartweed leaves

Pennsylvania smartweed plant

Ladysthumb ocrea contains hair

Pennsylvania smartweed

Leaf shape and arrangement—Leaves: Long and slender leaves arranged alternately along the stem, with a sheath at the base extending around the stem, leaf-shaped lanceolate to elliptic in outline, approximately 50−152 mm and 32 mm wide; leaves stalked on a short petiole; older leaves slightly hairy, leaves often but not always have a purple spot in the middle; the main distinguishing characteristic is a thin paper-like sheath called ocrea; the ocrea on Pennsylvania smartweed does not have hair, while the ocrea on ladysthumb has hair.

Stolon/Rhizome/Roots: No stolons or rhizomes; stem with swollen nodes and ocrea, stem may have reddish tint; fibrous roots system with shallow taproot.

Inflorescence: Flowers arranged in a spike-like cluster at the end of the stem, individual flowers are small and range in color from bright pink to rose to pink, five-parted.

Ladysthumb smartweed flowers *Source: Lachlan Cranswick*

Smartweed, Pennsylvania—*Persicaria pensylvanica* (L.) M. Gómez

Ladysthumb or Smartweed—*Polygonum persicaria* L. = *Persicaria maculosa* Gray

Other common names: Pinkweed, pink knotweed
Family: Polygonaceae
Life cycle: Warm-season, summer annual
Native to: United States

Distribution and Adaptation
- Distributed throughout the United States.

Morphology/Growth Pattern
- Pennsylvania smartweed is an annual plant that grows 1−4 ft (0.30−1.2 m) tall; the plant branches occasionally, semierect, often bending toward the light if growing in a partially shaded area; stems are round and light green to slightly red and has a tendency to zigzag between the short narrow petioles of the leaves; leaves are long (7 in. (18 cm) long and 3 in. (8 cm) across), green to dark green, shaped as lanceolate to broadly lanceolate with smooth margins with no hair on both upper and lower surfaces; leaves have some dark markings on the upper surface; a thin sheath called an ocrea wraps around the stem, ocrea have no hair; some stems terminate with a short spike-like flower head with small flowers ranging from dark pink to sometimes white.
- Reproduces by seed.

Use and Potential Problems
- No known commercial feed values. Weed of horticultural, agronomic, and nursery crops.

Toxicity
- None recorded.

Similar Species
- *Pennsylvania smartweed* and *ladysthumb* are similar in appearance and growth habit, however, ladysthumb has stiff hairs on the top edges of the ocrea unlike that of Pennsylvania smartweed where the ocrea are hairless; another similar plant to both Pennsylvania smartweed and ladysthumb is *tufted knotweed (Polygonum caespitosum* L. var *longisetum)*, this plant also has hair on the ocrea like ladysthumb, however, the hairs on the ocrea are much longer than those of ladysthumb.

Reference
- Smartweeds [14,17−19,24,33,46,47,58,61,66,67,72,76]

Sorrel, Red—*Rumex acetosella* L.

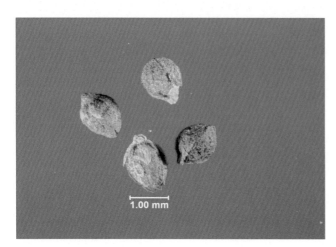

Red sorrel seeds

Red sorrel stem and ocrea *Source: James Altland*

Red sorrel flower/seedhead *Source: NCSU*

Seed

- Size: 1.2−1.5 mm long, 0.9−1.2 mm wide (with perianth attached); 1−1.3 mm long, 0.8−1.1 mm wide (without perianth), smaller than curly dock
- Shape and texture: triangular; perianth (papery covering) may be on seed; ovate with an extended base, three-angled, angles rounded but well defined; surface granular with network of veins, shiny without perianth
- Color: reddish-brown to dark brown

Red sorrel plant *Source: NCSU weeds*

Leaf shape and arrangement—Leaves: Young leaves are shaped like eggs and form a rosette at earlier stages and later alternate along the stem, older leaves develop basal lobes, with arrowhead-shaped appearance (two-pointed base lobes); a membranous sheath covers the stem at the base of the leaf.

Stolon/Rhizome/Roots: No stolons, rhizomatous, upright slender stem, branching above, ridged, often reddish; swollen node with ocrea; brownish red running rootstocks; taproot.

Inflorescence: Very small flowers are formed in raceme-like terminal clusters; male flowers are greenish-yellow, while female flowers are red to burnt orange, petals absent.

Sorrel, red—*Rumex acetosella* L.

Other common names: Sheep, field or mountain sorrel, sour dock
Family: Polygonaceae
Life cycle: Warm-season perennial
Native to: Europe

Distribution and Adaptation
- Distributed throughout the United States.
- Red sorrel is often found in wet and acidic soils of low fertility; competition from other plants on fertile soils reduces its abundance.

Morphology/Growth Pattern
- Male and female flowers are produced on different plants.
- Red sorrel is a rhizomatous, creeping perennial; the plant has slender, erect, simple branching stems arises from rosette of leaves and can reach 6–18 in. (15–46 cm) in height; flowers borne on the long raceme-like stalk at the end of the slender stem; plant is long and narrow, lower leaves are shaped like arrowhead with two small lobes at the base; creeps using an extensive shallow system of roots and rhizomes.
- Reproduces by seed and rhizomes.

Use and Potential Problems
- Red sorrel is primarily a weed of turfgrass, lawns, roadsides, landscapes, some cleared/burned areas, and some nursery crops; especially problematic in red clover fields grown for seed production, due to the fact that red sorrel seeds are difficult to separate from red clover seeds.

Toxicity
- Potential for poisoning of livestock; many *Rumex* species' toxicity is soluble oxalates.

Reference
- Red sorrel [12,14,17–19,24,33,46,47,58,61,66,67,72,76]

Sowthistle, Perennial—*Sonchus arvensis* L.

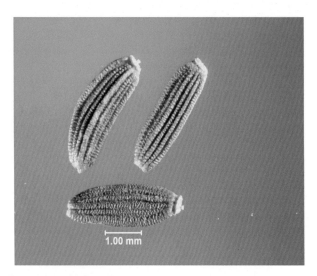

Perennial sowthistle seeds

Perennial sowthistle stem and leaves *Source: NCSU weed*

Perennial sowthistle flower *Source: John Wright*

Seed
- Size: 2.4−3 mm long, 0.7−1.1 mm wide
- Shape and texture: narrowly elliptical in outline, widest in the middle or slightly above gradually tapered to a truncate base, the truncate base and the apex have trimmed collars slightly curved; surface covered with prominent longitudinal ribs
- Color: dark brown to reddish-brown

Perennial sowthistle plant *Source: Peter Sforza*

Leaf shape and arrangement—Leaves: Alternate, lance-shaped, one leaf per node; lower leaves attach to the stem by way of winged petioles and have two to five deep lobes with soft prickly margins; upper leaves are less deeply lobed, with spiny-tipped teeth; leaves on middle and upper portion of the stem have shorter petioles; these leaves clasp the stem and have two to seven backward-pointing lobes per side; leaves on stems do not have petioles and have rounded lobes at the base; the majority of the leaves are found on the lower half of the stem; the shiny green leaves are 10−20 cm long.

Stolon/Rhizome/Roots: No stolons, rhizomatous; stems are smooth, hairless on lower part but upper parts on branches and flower stalks may have dark hairs, hollow, ridged and produces milky sap when injured; extensive downward, horizontally grown, numerous, fleshy, root systems.

Inflorescence: Flowers composed of many yellow ray flowers, similar to that of dandelion flowers but smaller and instead of being solitary, many heads are clustered at the ends of branched stems, 2.5−3.8 cm across.

Sowthistle, perennial—*Sonchus arvensis* L.

Other common names: Corn sow thistle, creeping sow thistle, field milk thistle, field sow thistle, milk thistle, swine thistle, tree sow thistle, marsh sowthistle
Family: Asteraceae/Compositae
Life cycle: Cool-season, perennial
Native to: Europe and western Asia

Distribution and Adaptation
- Found throughout the northern United States, southern Canada, and as far west as California.
- Inhibits disturbed areas, lawns, gardens, pastures, roadsides, waste places, no-tilled fields, and wetlands.
- Grows well on well-drained moist rich loamy soils.

Morphology/Growth Pattern
- Plant 2–4 ft (0.60–1.2 m) tall, forms loose patches from an extensive root system; succulent and all parts of the plant produces sticky milky sap when it is cut or injured, and has sour odor.
- Rosette leaves branching before flowering; leaves with clasping base, deeply lobed with spines on margins; upper leaves are fewer and much smaller than the basal leaves, flowers are bright yellow to yellow-orange in color; flower heads are few to several in numbers, large 1–2 in. (2.5–5 cm) across; deep yellow.
- Spreads by seeds and rhizomes.

Use and Potential Problems
- Considered noxious weed in most states.
- Although not preferred highly, plants are grazed by cattle and sheep.
- Known to be a heavy nitrogen user.

Toxicity
- None recorded.

Similar Species
- *Spiny sowthistle* (*Sonchus asper* L.) is very similar in appearance to *annual sowthistle* (*Sonchus oleraceus* L.), however, this species has rounded leaf lobes that clasp the stem, while annual sowthistle has distinctly pointed lobes. Additionally, the leaf margins of spiny sowthistle are much more spiny or prickly than those of annual sowthistle; both species arise from a taproot. *Perennial sowthistle* (*Sonchus arvensis* L.) arising from an extensive creeping root system has hollow stems, has no petiole; larger flower heads (than the annual sowthistles) and the lobes at the base of the leaves are clasping, small and rounded; the roots are horizontal (unique to perennial sowthistle), numerous, and fleshy but do not penetrate deeper than 4 in. (10 cm); the horizontal roots produce buds for new growth. Unlike other thistles the sowthistles produce milky sap when injured or broken. Another similar plant is prickly lettuce (*Lactuca canadensis* L.), the leaves of prickly lettuce are also lobed with spiny/prickly edges, however, their midribs have a row of stiff sharp prickles on the underside.

Reference
- Sowthistle, perennial [24,33,34,36,47,48,61,66,72,76]

Sowthistle, Spiny—*Sonchus asper* (L.) Hill

Spiny sowthistle seeds

Spiny sowthistle leaves and stem *Source: Peter Sforza*

Spiny sowthistle leaves and flower *Source: Lachlan Cranswick*

Seed

- Size: 2.7−3.0 mm long, 1.3−1.5 mm wide (without the pappus)
- Shape and texture: somewhat elliptical, gradually tapered to the base and narrow at the apex; surface dull, ridged, and wrinkled with a tuft of fine white hairs (white feathery pappus) attached to one terminal that collectively forms a white "puff ball" similar to that of dandelion
- Color: reddish-brown

Spiny sowthistle plant *Source: Peter Sforza*

Leaf shape and arrangement—Leaves: *Spiny/prickly sowthistle* are arranged in rosette, alternate on the stem, egg-shaped with toothed margins slightly prickly; leaves are glabrous and tend to fold upward along the central vein on the underside of each leaf; bluish-green in color.

Stolon/Rhizome/Roots: Spiny/prickly sowthistle have no stolons or rhizomes; have dull green or reddish green, round and smooth stems; have longitudinal veins; leaves and stems secrete milky sap when cut or injured; taproot.

Inflorescence: The upper stems of spiny sowthistle terminate in cluster of one to five composite on short stalks, each flower is about 15−17 mm across when fully open and consists of numerous yellow ray florets, the bases of each flower is short and is covered with dull green bracts.

Sowthistle, spiny/prickly—*Sonchus asper* (L.) Hill.

Other common names: Spiny annual sowthistle, spiny milk thistle, prickly sow thistle, sharp-fringed sowthistle
Family: Asteraceae/Compositae
Life cycle: Annual
Native to: Eurasia and tropical Africa

Distribution and Adaptation
- The sowthistles are found throughout the northern United States, and in parts of California, Texas, Missouri, and North Carolina.
- Grows on different kind of soils, including loam, clay loam, and shallow gravelly soils.
- Early invader of disturbed soils.
- All sowthistle species prefer moist rather than dry conditions, and do better under alkaline or neutral conditions.

Morphology/Growth Pattern
- An annual that may reach as much as 1−6 ft (0.30−1.80 m) tall in height (height very variable).
- Annual sowthistle branches sparingly in the upper half of the plant and the stems are green or reddish green, round and smooth; leaves have very prickly margins that initially develops from a basal rosette and then occur alternately on the flowering stem; the leaves are bluish-green in color; the yellow flowers have "puff-ball" seedheads.
- Annual sowthistles propagate by seed while perennial sowthistle spreads by seed and rhizomes.

Use and Potential Problems
- Primarily a weed of landscapes, small grains, pastures, hay fields, orchards, and road sides.

Toxicity
- None recorded.

Similar Species
- *Spiny sowthistle* (*Sonchus asper* L.) is very similar in appearance to *annual sowthistle*, however, this species has rounded leaf lobes that clasp the stem, while annual sowthistle has distinctly pointed lobes. Additionally, the leaf margins of spiny sowthistle are much more spiny or prickly than those of annual sowthistle; both species arise from a taproot. *Perennial sowthistle* (*Sonchus arvensis* L.) arising from an extensive creeping root system has hollow stems, has no petiole; has larger flower heads (than the annual sowthistles) and the lobes at the base of the leaves are clasping, small, and rounded; the roots are horizontal (unique to perennial sowthistle), numerous, and fleshy but do not penetrate deeper than 4 in. (10 cm); the horizontal roots produce buds for new growth. Unlike other thistles the sowthistles produce milky sap when injured or broken. Another similar plant is *prickly lettuce* (*Lactuca canadensis* L.), the leaves of prickly lettuce are also lobed with spiny/prickly edges, however, their midribs have a row of stiff sharp prickles on the underside.

Reference
- Spiny sowthistle [6,26,36,44,46−48,61,66,72,76]

*Speedwell, Corn—*Veronica arvensis* L.; Speedwell, Heath—*Veronica officinalis* L.; Speedwell, Persian—*Veronica persica* Poir.

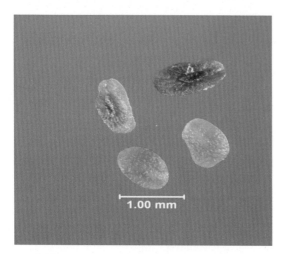

Corn speedwell

Corn speedwell stem and leaves *Source: NCSU weed*

Corn speedwell flowers *Source: NCSU weed*

Seed (Corn Speedwell)
- Size: 0.9—1.2 mm long, 0.6—0.7 mm wide
- Shape and texture: ovate with a notch at the small end, hilum raised in the center of the concave face; surface covered with fine wrinkles visible under high magnification
- Color: pale dull orange, somewhat translucent, hilum is darker

Corn speedwell plant *Source: NCSU weed*

Leaf shape and arrangement—Leaves: Lower leaves are round to egg-shaped, toothed on the margins with prominent veins, the upper leaves are linear in shape and are smaller and narrower than lower leaves; lower leaves are arranged opposite from each other while upper leaves alternate; the lower leaves are petioled, while the middle and the upper leaves are sessile.

Stolon/Rhizome/Roots: No stolons or rhizomes; stems erect, sprawling or decumbent.

Inflorescence: Flowers tiny (2—4 mm wide) deep blue-purple, found nested in the leaf axils on the upper part of the erect flowering stem and are on short stalks, four petals united at the base to form a short tube; fruits are flat, hairy, deeply notched on the top and rounded or pointed on the opposite, with the line of hair on the outer edge, and heart-shaped; each pod contains between 15 and 20 tiny yellow seeds.

*Plant description for corn speedwell unless indicated otherwise.

*Speedwell, corn—Veronica arvensis L.
Speedwell, heath—Veronica officinalis L.
Speedwell, Persian—Veronica persica Poir.

Corn speedwell *Source: NCSU weeds*

Corn speedwell *Source: NCSU weeds*

Heath speedwell

Heath speedwell

Persian speedwell *Source: NCSU*

Persian speedwell *Source: NCSU*

Speedwell, corn—*Veronica arvensis* L.

Other common names: Speedwell, rock speedwell, wall speedwell
Family: Scrophulariaceae
Life cycle: Winter annual

Speedwell, heath—*Veronica officinalis* L.

Other common names: Common speedwell, gypsyweed
Life cycle: Perennial

Speedwell, Persian—*Veronica persica* Poir.

Other common names: Bird's eye
Life cycle: winter annual
Native to: Europe and Asia

Distribution and Adaptation
- Found throughout the United States except the Rocky Mountains.
- Found in open sunny areas and grows on many soil types, often found in gravely, sandy poor soils, best adapted to dry or sandy soils and shaded areas.

Morphology/Growth Pattern
- Corn speedwell is the most common *Veronica* species.
    Low growing 2—6 in. (5—15 cm), succulent with ascending or prostrate stems that rise from the base of the plant. The leaves on the lower and middle stem are opposite and sessile while the lower leaves have short petioles, leaves oval to ovate in shape and are created along the margins; plant branches into many stems near the base of the plant, the stems typically are round, hairy, and fleshy; both the stems and the leaves have fleshy texture and scattered across the surface of the plant; has small blue flowers.
- Reproduces only by seed.

Use and Potential Problems
- Weed mostly found in highly disturbed areas such as edge of roads, fields, and cultivated fields.
- Often found in open rocky woods, waste grounds, cultivated areas, pastures, lawns, roadsides, and railroads.

Toxicity
- None recorded.

Similar Species
- Similar species to corn speedwell include *Persian speedwell* and *heath/common speedwell*. Persian speedwell and corn speedwell are annuals while heath speedwell is perennial. The leaf shape of Persian speedwell and corn speedwell are similar, however, the leaves of *Persian speedwell* are much larger, are deeply lobed, and are less hairy than those of *corn speedwell*, the lower leaves of both *Persian speedwell* and *corn speedwell* are arranged opposite from each other while the upper leaves alternate on the stem. Additionally, the size of the stem of corn speedwell is progressively reduced upward while the leaf size of *Persian speedwell* remains the same on vegetative as well as flowering stems.

Reference
- Corn speedwell [14,24,33,44,47,48,72]

Spurge, Spotted—*Chamaesyce maculata* (L.) Small; Spurge, Prostrate—*Euphorbia humistrata* Engelm. Ex Gray

Spotted spurge seeds

Spotted spurge hairy stems, spotted leaves, and glowers

Spotted spurge leaves, stem, and flowers *Source: NCSU weed*

Seed

- Size: 0.8−0.9 mm long, 0.6 mm wide
- Shape and texture: oblong with one end tapered to a blunt tip, form oblong, four-angled, four-sided, narrow dark line between the two smaller faces; surface with large faint crosswise wrinkles, dull whitish outer layer
- Color: reddish brown with a grayish or whitish cast

Spotted spurge: stem, leaves, and flowers *Source: NCSU weed*

Leaf shape and arrangement—Leaves: Egg-shaped, 4−15 mm long, and opposite; leaves on very short petioles, leaf margins may be finely toothed near the leaf apex; mostly no hair but occasionally can find long hairs, leaves have maroon spots on the upper surface, variable leaf size 4−15 mm long.

Stolon/Rhizome/Roots: No stolons or rhizomes; stems are prostrate, branching from a central point, densely hairy and pink to reddish in color; small taproot and many fibrous root systems.

Inflorescence: Flowers arise at the leaf axils and are white or greenish.

Spurge, spotted—*Euphorbia maculata* (L.) Small

Spurge, prostrate—*Euphorbia humistrata* Engelm. Ex Gray

Other common names: Milk spurge, prostrate spurge, spreading spurge
Family: Euphorbiaceae
Life cycle: Summer annual
Native to: Europe

Distribution and Adaptation
- Found throughout the Eastern part of the United States.

Morphology/Growth Pattern
- Erect or prostrate growth habit sending out branches 4−12 in. (10−30 cm) long in all directions.
- Spotted spurge leaves are opposite, elongated with purple spots on the surface, the leaf margins have small teeth; the plant produces a milky sap when injured or broken; the plant can form a dense mat that rises from a central point; the cup-shaped flowers are white or greenish.
- *The stems of spotted spurges do not root at the nodes.*
- Reproduces by seed.

Use and Potential Problems
- Primarily weed of disturbed areas, lawns, pastures, and waste areas.
- All plant parts are poisonous; when ingested cause nausea, vomiting, and diarrhea, may also cause skin irritation.

Toxicity
- Diterpene euphorbol esters, esulones, miliamines.

Similar Species
- In some cases, *prostrate spurge* (*Euphorbia humistrata* Engelm: ex Gray) and *spotted spurge* are described as the same species. However, among other differences between these species, *prostrate spurge* roots at the nodes while *spotted spurge* does not. Other distinguishing characteristics of prostrate spurge from spotted spurge include that prostrate spurge leaves are pale green, egg-shaped and approximately half length of the leaf is cleft. Other plant may be confused with the spurges is *prostrate knotweed* (*Polygonum aviculare* L.). Although prostrate knotweed has a prostrate growth habit and forms a dense mat like *prostrate spurge*, prostrate knotweed, as other members of its genus has an ocrea (peppery cover) on the stem node while the spurges do not; *prostrate knotweed* also does not produce milky sap like the spurges.

Reference
- Spotted spurge [12,14,17,24,33,36,44,48,62,67,72,74,76]

Star-of-Bethlehem—*Ornithogalum umbellatum* L.

Star-of-Bethlehem plant *Source: Jeff Pippen*

Star-of-Bethlehem flower *Source Jeff Pippen*

Leaf shape and arrangement—Leaves: Basal, grass-like, narrow, linear, equal or longer than flowering stalks, pale-green to white, mid-veins light-colored.

Stolon/Rhizome/Roots: No stolons or rhizomes; have underground bulbs.

Inflorescence: Flowers six slender white petals with a characteristic green strip on underside; the flowers grow in an umbel-like cluster on top of yellowish-green stalks that branch out alternately from the main stalk; seed capsules three-angled; seeds are black and few seeds/compartments.

Star-of-Bethlehem—*Ornithogalum umbellatum* L.

Family: Liliaceae
Other common names: Sleepydick, summer snowflake, star-flower, the Pyrenees, sand lily, mountain lily, sage lily, wild tuberose, star lily, summer snow-flake
Life cycle: Cool-season, bulbous perennial
Native to: Europe, Asia and North Africa and the Middle East

Distribution and Adaptation
- Found primarily in the Piedmont region of the southern states.

Morphology/Growth Pattern
- The leaves of star-of-Bethlehem are grass-like and are enclosed by membranous white sheath at the base, slightly folded, flowers are white, showy, in an umbel-like cluster; flowering stem reaching 30 cm in height and is leafless; plant arising from a stout rootstalk with fleshy fibrous roots.
- Perenniate from a central bulb and by seeds.

Use and Potential Problems
- Weed in turfgrass, landscapes, and old flower gardens.
- Plant matures early in season and thus often escapes most weed control measures including cultivation and chemical means.
- All parts of this plant are poisonous and consumption by livestock can result in death.
- Sold commercially as a spring flowering ornamental.

Toxicity
- Cardenolides (Table 3I).

Similar Species

Reference
- Star-of-Bethlehem [14,24,39,47,48,72]

St. John's Wort, Common—*Hypericum perforatum* L.;
St. John's Wort, Spotted—*Hypericum punctatum* L.

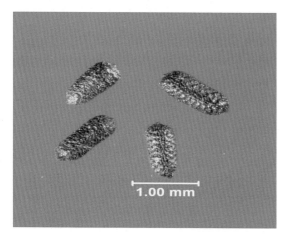

St. John's wort seeds

St. John's wort leaves *Source: James Altland*

St. John's wort leaves and stem *Source: James Altland*

St. John's wort flower *Source: James Altland*

Seed
- Size: 1.0—1.3 mm long, 0.4—0.5 mm wide
- Shape and texture: oblong in outline, circular in cross section; apex rounded with short fine tip, some seed curved; surface roughened with a fine network of veins; slightly glossy appearance
- Color: dark brown to black

St. John's wort plant *Source: James Altland*

Leaf shape and arrangement—Leaves: Arranged opposite and are small, rather oblong but broader at the sessile base than at the tips of the leaves; the leaf blade when held up to a light source, has translucent dots, leaf margins are entire (not toothed) and may have black dotted.

Stolon/Rhizome/Roots: No stolons, extensively rhizomatous.

Inflorescence: Numerous flowers clustered at the end of stems or branches; 1.7 cm in diameter and have black-dotted yellow petals; the stamens are numerous and appear in three clusters. Flowers are yellow or orange-yellow with four to five petals and sepals.

St. John's wort—*Hypericum perforatum* L.

Other common names: Klamathweed, penny-Jon, Rosin, Rosin-Rose, speckled John, John's wort, Herb-John, Tipton weed, goat weed, Eola weed
Family: Hypericaceae
Life cycle: Cool-season, perennial
Native to: Europe

Distribution and Adaptation
- In the United States found in the Pacific Northwest
- Well adapted to sandy or gravelly soils.

Morphology/Growth Pattern
- There are over 75 species of *Hypericum* in the Southeast of the United States.
- The tough, branching, smooth stems arise from the crown of a much branched root-stock; has an extensive rhizome, up to 1 m tall, typically growing in patches; stems are 2–3 ft (0.60–0.90 m) high; erect and highly branched, rust colored, woody at the base; leaves opposite, sessile, entire, elliptic to oblong, covered with transparent dots; flowers are 0.5 in. (1.7 cm) in diameter bright yellow, numerous in flat topped cymes, with five separate petals with occasional minute dots around the edges, petals are twice as long as the sepals, stamens are numerous; seedpods are a ¼ in. (0.6 cm) long, rust-brown, three-celled capsules, each with numerous seeds.

Use and Potential Problems
- Causes a serious problem in rangeland; not only the plant is an aggressive invader, it is also toxic to grazing animals.
- Causes a serious skin disorder called photosensitization, light skinned animals are particularly susceptible.
- Contains a toxic substance that affects white-haired animals. Affected animals rarely die but often lose weight and develop skin irritation when exposed to strong sunlight.

Toxicity
- Photosensitization agents.

Reference
- St. John's wort [12,17,19,33,46,47,74,76]

Stickweed or Yellow Crownbeard—*Verbesina occidentalis* (L.) Walter

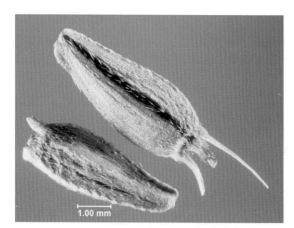

Stickweed seed

Stickweed stem with opposite leaves *Source: Jeff Pippen*

Stickweed flowers *Source: Jeff Pippen*

Seed
- Shape and texture: long, with hooks on the broader side of the seed, rough surface
- Color: light to dark brown

Stickweed field

Leaf shape and arrangement—Leaves: Opposite, lanceolate to ovate, 8−20 cm long, 5−10 cm wide, tapering at both ends, the base of the leaf projects as a weak wing on the stem; leaves are smooth (no hair), margins toothed or serrated.

Stolon/Rhizomes/Roots: No stolons or rhizomes; stem erect, smooth with occasional hairs, often not branched, the stem has "wings" that run the length of the entire plant; has large perennial basal crown from which new growth arises.

Inflorescence: Flowers appear in clusters at the end of the erect stems; each flower consists of yellow rays, 1−2 cm long, 4−7 mm wide, notched or entire at apex, and inner small yellow-green disk, as the plant matures, the ray florets droop downward.

Stickweed or Yellow Crownbeard—*Verbesina occidentalis* (L.) Walter

Other common names: Yellow crownbeard
Family: Asteraceae/Compositae
Life cycle: Cool-season, perennial
Native to: Europe

Distribution and Adaptation
- Grows on a variety of open to semiopen areas, often found in disturbed areas such as cleared wood lands, pastures, fields, roadsides, and waste ground.

Morphology/Growth Pattern
- Tall growing perennial plant that grows 6–10 ft (1.8–3 m) in height more; leaves on the plant are arranged opposite from each other and are shaped lanceolate, "winged" stem that runs along the entire length of the stem, showy yellow flowers.
- Propagated by seed.

Use and Potential Problems
- Primarily a weed in pastures, hay fields, woodland, fencerows, woodlands, and right-of-way.

Toxicity
- None recorded.

Similar Species
- *Stickweed* is often confused with the more *common wingstem* (*Verbesina alternifolia*), both species occupy similar habitat and flower at the same time. Some of the characteristics that separate the two species are leaf arrangements, i.e., *wingstem* has alternating leaves while *stickweed* leaves are arranged oppositely along the stem.

Reference
- Stickweed or Yellow Crownbeard [48,58]

Teasel, Common—*Dipsacus fullonum* L.

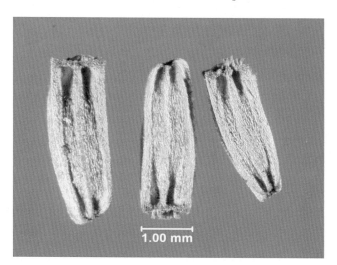

Common teasel seeds

Common teasel leaf—short prickles on the midrib *Source: Jams Altland*

Common teasel stem

Seed
- Size: 3.0–4.0 mm long, 0.9–1.2 mm wide
- Shape and texture: nearly rectangular or oblong in outline, four-angled; surface ribbed appearance and are covered with very fine straight white hairs over the entire surface, rigid
- Color: light brown with pale ridges

Common teasel plant

Leaf shape and arrangement—Leaves: Basal leaves are widest in the upper part of the plant and tapering to the base, with rounded teeth along the leaf margins and prickly, 20–60 cm long, leaves have a wrinkle appurtenance; leaves on flowering stem are lanceolate, arranged opposite, sessile (no petiole) and have clasping leaf base completely surrounding the stem; short prickles on the leaf lower midrib.

Stolon/Rhizome/Roots: No stolons or rhizomes; stems are erect and branching on upper part of the plant, stem is striate-angled with several rows of downward turned numerous prickles; shallow thick taproot with secondary fibrous root system.

Inflorescence: The flowers are in dense, spiny heads with long, slender, stiff, bracts underneath them; flowers bloom in the circular matter around the seedhead; individual flowers are small, and are 10–15 mm long and have white petals that are united into a tube with four purple lobes, occur on flower stalks (peduncles); several long leaf-like bracts also branch out from the bases of the flower and curve upward around the head.

Teasel, common—*Dipsacus fullonum* L.

Other common names: *Dipsacus sylvestris* L., barber's brush, brushes and combs, card teasel, card-thistle, church broom, gypsy-combs, Venus' basin, Venus'-cup, wild teasel, water thistle
Family: Dipsacaceae
Life cycle: Cool-season, biennial
Native to: Europe and North America

Distribution and Adaptation
- Found throughout the United States except the northern great plains. Most common in the northeastern and Pacific Coast States
- Prefers damp, coarse, and fertile soils.
- Commonly found in pastures, abandoned fields, roadsides, railroads, and waste areas.

Morphology/Growth Pattern
- The plant initially produces a basal shiny rosette leaves and flowering stems are produced in the second year.
- The plant is an erect biennial that reaches up to 7 ft (2 m) tall; leaves on the stem are arranged oppositely, are large, oblong, and have wrinkled appearance with scallop margin, the two opposite leaf bases fuse together to form a "cup" like structure that commonly collects rainwater; stem has small prickles that curve downward; the characteristics large spiny flower heads are showy, distinct, stiff, and spiny. Common teasel reproduces by seed.

Use and Potential Problem
- Although considered a weed it is often used in flower arrangements.
- Common thistle can be controlled by cultivation (does not survive disturbances).

Toxicity
- None recorded.

Similar Species
- Similar species include *cutleaf teasel* (*Dipsacus laciniatus* L.), however, cutleaf teasel can be distinguished from common teasel by its deeply lobed leaves; additionally cutleaf teasel usually has white flowers, while common teasel usually has purple flowers. Common teasel at a rosette stage can be confused with thistle, *common burdock* (*Arctium minus* L.) and *broadleaf duck* (*Rumex obtusifolius* L.); however, none of these plants have "wrinkled" leaves like those of common teasel.

Reference
- Common teasel [17,19,24,33,47,48,66,72,74,76]

Thistle, Bull—*Cirsium vulgare* (Savi) Ten.

Bull thistle seeds

Bull thistle leaves *Source: Peter Sforza*

Bull thistle *Source: Peter Sforza*

Seed

- Size: 3.5−4.0 mm long, 1.2−1.7 mm wide
- Shape and texture: elongated, broadest above the middle of the seed, gradually tapered to a narrow, truncate base, slightly depressed, swollen near the apex, apex tipped with a parachute of bristle which is often broken off; surface smooth, dull or with a slight sheen
- Color: dull orange-brown to pale gray-brown, collar is paler, and there are a few dark, narrow lengthwise streaks

Bull thistle plant *Source: Fred Fishel*

Leaf shape and arrangement—Leaves: The rosette leaves are large, deeply lobed, coarsely toothed and with sharp, marginal prickles; the upper surface of the leaves has short prickles, and the lower surface is covered with white hairs; the leaves on the stem are similar to the rosette leaves but are smaller.

Stolon/Rhizome/Roots: No stolons or rhizomes; the flowering stem is about 1.2 m high with prickly wings; thick, fleshy taproot the first year, with the secondary rooting system the second year.

Inflorescence: The flower heads have numerous, spine-tipped bracts; the heads are up to 5 cm in diameter; the flowers are all tubular and purplish; the hairs on the fruit are feather-like; disk florets are deep purple, whitish.

Thistle, bull—*Cirsium vulgare* (Savi) Ten.

Synonyms: *Carduus vulgare* L., *Cirsium lanceolatum* L.
Other common names: Spear thistle, spear thistle, bur thistle, plume thistle, lance-leaved thistle
Family: Asteraceae
Life cycle: Summer biennial
Native to: Europe and Asia

Distribution and Adaptation
- Wide spread throughout the United States.
- Prefers fertile and moist soils.
- Found in pastures, meadows, disturbed waste places, and other uncultivated ground.

Morphology/Growth Pattern
- A biennial taprooted plant, spiny stems winged especially in the upper portion of the plant.
- Leaves on the first year for a rosette, the stem leaves are lobbed, basal and lower leaves larger than the leaves on the upper part of the plant; leaves have long sharp pines, the upper surface of the leaves has short prickles while the underside of the leaves is cottony.
- Can grow 2–5 ft (0.60–1.5 m) with many spreading branches.

Use and Potential Problem
- This highly competitive plant has no commercial value.
- Considered a weed in all forage and row crops as well as gardens, landscape, and roadsides; serious weed in pastures, avoided by grazing animals.

Toxicity
- None recorded.

Similar Species
- Bull thistle is often confused with *Canada thistle* (*Cirsium arvense* L.). However, Canada thistle is a rhizomatous plant with much smaller flowers, glabrous with weak prickles and no spines on the leaves. The short yellowish spines on the upper surface of the leaves will help distinguish this thistle from all others.

Reference
- Bull thistle [17,19,24,33,39,47,48,66,72,74,76]

Thistle, Canada—*Cirsium arvense* (L.) Scop.

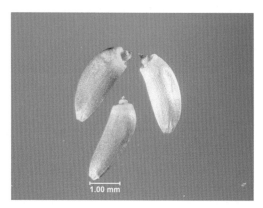

Canada thistle seeds

Canada thistle leaves

Canada thistle flower head

Seed
- Size: 2.5−3.0 mm long, 0.8−1.0 mm wide
- Shape and texture: elongated, broadest above the middle of the seed toward top, gradually tapered to a narrow, truncate, base; truncate apex, apex usually with style and tipped with a parachute of bristles which often easily break off; surface smooth, dull, glossy
- Color: light brown, the collar is yellow

Canada thistle plant

Leaf shape and arrangement—Leaves: Alternate on the stem lacking petiole, oblong and 5−15 cm long, leaves without petiole have curled wavy surface, several prickly toothed, irregularly shaped lobes, the underside of the leaves are often covered with soft, wooly words.

Stolon/Rhizome/Roots: No stolons, creeping, extensive rhizomes; stem grooved, rigid and prickly; fibrous root system.

Inflorescence: Flower heads are global or terminal, usually branched inflorescence; composed of numerous pinkish purple occasionally white disc florets; the flower head is surrounded by overlapping spiny floral bracts; the male and female flower heads appear on different plants, male flower heads are oblong, while female flower heads are flask-like; each female flora produces a single seed; no fruit produced by the male plant.

Thistle, Canada—*Cirsium arvense* (L.) Scop.

Other common names: Creeping thistle, field thistle, cursed thistle, corn thistle, small- flowered thistle, green thistle, perennial thistle, soft field thistle, green thistle
Family: Asteraceae
Life cycle: Warm-season, perennial
Native to: Europe

Distribution and Adaptation
- Found throughout the northern half of the United States and Canada.
- Mostly found in wet or dry abandoned sites such as roadsides, burned areas, slash piles, shores, and stream margins. Soil disturbance appears to ensure its establishment.

Morphology/Growth Pattern
- Leaves are oblong, alternate, toothed irregularly lobed, spines on the leaf margins or rarely almost spineless; the leafy and hollow stem can grow up to 1.6−4.9 ft. (0.50−1.5 m) tall, the stem becomes hairy with age.
- Canada thistle is a colony-forming perennial deep and extensive rhizomes.
- The deep rhizomes allow the plant to survive below the cultivation zone.
- Approximately eight buds per meter of rhizome and 1 year old plant may have up to 200 buds. A single plant is capable of producing over 20 ft (6 m) of rhizome.
- Reproduces by seed and creeping rhizomes.

Use and Potential Problem
- No commercial value, considered an invasive weed.
- Difficult to control, breaking up the rhizomes by plowing only increases the number of plants.

Toxicity
- Nitrate.

Similar Species
- Canada thistle is different from the other thistles in that they have male and female flower heads on a different plant.

Reference
- Canada thistle [12,17,19,24,33,39,47,48,61,66,72,74]

Thistle, Musk—*Carduus nutans* L.

Musk thistle seeds

Musk thistle leaves and stem *Source: Jeff Pippen*

Different stages of bull thistle flower heads

Seed
- Size: 3.4−4.5 mm long, 1.2−1.7 mm wide
- Shape and texture: elongated, curved, sometime faintly four- or five-sided, apex truncate and has a distinct thin, slightly flaring collar; surface smooth, very shiny, many slight lengthwise ridges
- Color: yellowish-brown, yellow band at the apex

Musk thistle plant

Leaf shape and arrangement—Leaves: Arranged in basal rosette or 6−8 leaves in the first year, these leaves are up to 30 cm long and 15 cm wide, the basal leaves extend down to the stem with spiny wing appearance, leaves deeply to shallowly lobed and irregularly spiny leaf margins; both surfaces of the leaves covered with soft, wooly hairs; light green leaves with white midrib.

Stolon/Rhizome/Roots: No stolons or rhizomes; stems stout, erect, spiny, branched; long thick fleshy taproot.

Inflorescence: Flower heads are global, solitary 3.8−7.6 cm in diameter and usually bent over, flowers are deep rose, violet, or purple; flowers are subtended by broad spine-tipped bracts yellowish in color; fruits are smooth, yellowish brown with plume of white hair-like fibers, striated, 0.5 cm long, shiny, yellowish-brown with plume of white hair-like fibers.

Thistle, musk—*Carduus nutans* L.

Other common names: Nodding thistle, plumeless thistle
Family: Asteraceae
Life cycle: Winter annual or biennial
Native to: southern Europe and western Asia

Distribution and Adaptation
- Distributed throughout the United States.
- Invades pastures, roadsides, forest lands waste areas ditch banks, stream banks, and grain fields.

Morphology/Growth Pattern
- Leaves alternate on the stem and are smooth, dark greenish, a light green midrib and whitish margin; leaves are deeply dissected, each leaf with five spines at the tip. Flowers with spine-tipped bracts, deep pink to purple and often have nodding heads. Plant produces musk sent, hence called musk thistle.
- Grows up to 6 ft (1.8 m) tall.
- Reproduces by seed.

Use and Potential Problem
- Considered as one of the most troublesome weeds in pastures, nursery crops, waste areas, roadsides, and ditches.

Toxicity
- Nitrate.

Similar Species
- Similar species include *bull thistle* (*Cirsium vulgare* (Savi) Tenore). Bull thistle has rough hairs on the upper surface of the leaf blade and whitish softer hair below, while musk thistle leaves are mostly hairless; additionally, the flowering heads of bull thistle have a constriction at the junction of the bracts and the flowers, whereas musk thistle does not have such constriction between the bracts and flower head. *Canada thistle* (*Cirsium arvense* (L.) Scop) can be confused with musk thistle. However, Canada thistle is a rhizomatous perennial and its leaves not as deeply lobed as musk thistle.

Reference
- Musk thistle [12,17,19,24,33,44,47,48,61,66,72,74,76]

Velvetleaf—*Abutilon theophrasti* Medik.

Velvetleaf seeds

Velvetleaf leaf

Velvet leaf seedpod

Source: Fred Fishel

Seed

- Size: 3.0−3.6 mm long, 2.2−2.6 mm wide
- Shape and texture: kidney-shaped or mitt-shaped, notched on one side; thickest along the outer margin; seed surface covered with sparse short hairs along inner margins of lobes, and fungus-like growth on seed surface; seed generally appears netted and rough
- Color: grayish brown

Velvetleaf plant

Leaf shape and arrangement—Leaves: Alternate, large, heart-shaped, 5−20 cm wide, leaves sharply pointed at the apex; the leaves are covered with soft, velvety hairs; leaf margins are entire or irregularly toothed and are hairy.

Stolon/Rhizome/Roots: No stolons or rhizomes; stem thick, sparingly branched, covered with star-shaped hairs; taprooted.

Inflorescence: Flowers born on short peduncles shorter than petiole; composed of five sepals, five petals and numerous stamens that are united to form a column around the style; orangish yellow flowers, 15−20 mm across, appears in clusters or single in the axils of leaves; large characteristic seed pods originate from leaf axis; seedpods are green and turn black at maturity; seedpods like chopped off okra pod with distinct points, generally black or dark gray.

Velvetleaf—*Abutilon theophrasti* Medik.

Other common names: Indian mallow, butter print, velvet-weed, Indian hemp, cotton-weed, button-weed, pie-maker, elephant ear
Family: Malvaceae
Life cycle: Warm-season, summer annual
Native to: Asia/India

Distribution and Adaptation
- Widely distributed in the central United States and Canada.
- Thrives on rich soils and is found in cultivated fields, pastures, gardens, fence rows, and waste areas.

Morphology/Growth Pattern
- Completely covered with fine hairs, erect stem, branched, 2−7 ft (0.60−2 m) tall.
- Leaves alternate, heart-shaped, pointed at the apex, 5 in. (13 cm) or more in width on a slender petioles; flowers are solitary in leaf axis with five yellow-orange petals and numerous fused stamens that form a tube.
- Propagated by seeds and the seeds can remain viable in the soil for more than 50 years.

Use and Potential Problem
- Mainly a weed in pastures, roadsides, and cultivated fields.
- Due to the fact that the seed remains for many years, velvetleaf is hard to eradicate.
- Used as a fiber plant in some parts of the world.

Toxicity
- None recorded.

Reference
- Velvetleaf [6,17,19,24,33,36,47,48,61,67,74,76]

Venus Lookingglass, Common—*Specularia perfoliata* L. Nieuwl., Formerly Known as *Triodanis perfoliata* (L.) [A. DC.]

Venus lookingglass stem and leaf

Common Venus lookingglass plant

Venus lookingglass flowers

Leaf shape and arrangement—Leaves: Alternate, clasping, and surrounding the stem; upper leaves usually smaller than the leaves on the lower part of the stem, coarsely serrate to crenate, to 2 cm long, broadly ovate, glabrous on the upper surface.

Stolon/Rhizome/Roots: No stolons or rhizomes.

Inflorescence: Each flower sits above a clasping leaf on a wand-like stem; flowers have five regular parts and are blue sometimes purple in color; only a few flowers will open at any one time.

Venus lookingglass, common—*Specularia perfoliata* L. Nieuwl., formerly known as *Triodanis perfoliata* (L.) [A. DC.]

Other common names: Clasping Venus' Looking-glass
Family: Campanulaceae
Life cycle: Warm-season, winter annual
Native to: North America

Distribution and Adaptation
- Native to the United States. Found in the most part of the United States except for the Rocky Mountains.
- Found usually in infertile soils, dry fields, fencerows, and open woods.

Morphology/Growth Pattern
- Annual herbaceous plant that can grow to a height of 30 in. (75 cm) but usually much shorter than that.
- Common Venus lookingglass has an erect to ascending stems; the leaves on the plant alternate each leaf toothed and clasping on the stem; stems are usually branched at the base; bluish-violet tubular flowers are parted into five regular parts.
- Reproduces by seed.

Use and Potential Problem
- Considered a weed.

Toxicity
- None recorded.

Reference
- Common Venus lookingglass [14, 44, 48, 58, 72, 76]

Vervain, Blue—*Verbena hastata* L.

Blue vervain seeds *Source: Image http://www.psu.missouri.edu/fishel*

Blue vervain leaves at the stem

Blue vervain flowers *Source: John Wright*

Seed
- Size: 1.7–2.0 mm long, 0.7–0.9 mm wide; thickness 0.4–0.5 mm
- Shape and texture: oblong, both ends bluntly rounded, longitudinal ridge where two faces meet; surface nearly smooth, somewhat shiny, on the convex side there are three faint lengthwise ridges as well as several wrinkles near the apex; flat side of the seed my appear scurfy, with scattered white flecks
- Color: orange-brown with translucent marginal wing

Blue vervain plant *Source: John Wright*

Leaf shape and arrangement—Leaves: Lance-shaped/lanceolate to ovate leaves, simple, arranged opposite from each other or whorled, 18 cm long and 2.5 cm wide; leaf edges are toothed or serrated; short petioled and are prominently pinnately veined; leaves are green on the upper surface and have a grayish-green pubescence on the lower surface.

Stolon/Rhizome/Roots: No stolons or rhizomes; stem sometimes branches but often simple from the base, square or four-sided, grows from thick root stalk.

Inflorescence: Flowers few to numerous, dense, straight terminal spikes arranged in upright panicles, blue to purplish or pinkish, five petals, lavender, blue, violet, pink or occasional blue, and rarely white.

Vervain, blue—*Verbena hastata* L

Other common names: Verbain, wild vervain, wild hyssop, wolly verbena, false verbain, simpler's-joy, ironweed, American blue vervain, herb of grace, swamp verbena
Family: Verbenaceae
Life cycle: Warm-season perennial
Native to: North America

Distribution and Adaptation
- Grows throughout the United States and Canada.
- Occurs in moist fields, roadsides, waste places, damp thickets, shores, near springs and areas of seepage, bogs, marshes, and along stream banks.
- Prefers gravelly or heavy loam soils.
- Shows remarkable resistance to drought and is one of the most common invaders of run-down, dried-out pastures and ranges.
- Spreads fast, forming almost pure colonies or stands of considerable areas.

Morphology/Growth Pattern
- Perennial, erect, often branching forb that grows 2–7 ft (0.6–2 m) tall; square stem with rough hair; leaves broadly lance shaped and sharply toothed; produces numerous to few blue-violet sometimes white flowers arranged on slender terminal spikes.
- Roots can grow to depth of 4–5 ft (1.2–1.5 m).
- Reproduces by seed and short rootstocks.

Use and Potential Problems
- Due to its bitter taste, animals do not graze the plant.
- This herb reportedly has been used by Native Americans as both food and medicine; used in treatments of colds, coughs, gastrointestinal disorders, and stomach aches; seeds used for food when roasted and ground into powder.
- Seeds are consumed in a small amount by Northern Cardinals and sparrows; flowers are attractive to numerous species of butterflies.

Toxicity
- None recorded.

Similar Species
- Thirteen species of *Verbena* are found in the southeast; other similar species include *Verbena brasiliensis* Vell (Brazil vervain), *Verbena rigida* Spreng (rigid vervain), and woody verbena (*Verbena stricta* L.).

Reference
- Blue vervain [18,34,39,57,58]

Virginia Copperleaf—*Acalypha virginica* L.

Virginia copperleaf seeds

Virginia copperleaf leaves

Virginia copperleaf stem and flowers

Seed
- Size: greater than 1 mm in length
- Shape and texture: pear-shaped and slightly
- Color: dull reddish-brown to gray with reddish-brown spots, mottled

Virginia copperleaf plant

Leaf shape and arrangement—Leaves: Egg-shaped, lower leaves arranged opposite on the stem, while subsequent upper leaves alternate along the hairy stems; leaves on petiole with petiole exceeding the floral bracts in the axil, **leaf blade**: ovate, shallowly cleft 9—16 lobes longer than its petiole.

Stolon/Rhizome/Roots: No stolons or rhizomes; stems hairy, 30—60 cm tall; root shallow with secondary fibrous root system.

Inflorescence: Small flowers clustered in the leaf axis (between stem and leaf petiole); greenish in color; seedpod in three-seeded surrounded with deeply cut leaf-like structure.

Virginia copperleaf—*Acalypha virginica* L.

Other common names: Three-sided Mercury
Family: Euphorbiaceae
Life cycle: Summer annual
Native to: North America

Distribution and Adaptation
- Widely distributed Maine to Indiana and Iowa, south to Florida and Texas.
- Found in dry or moist open woods, fields, and road sides.

Morphology/Growth Pattern
- The plant is erect and branching stems are hairy to moderately hairy, 1—2 ft (30—60 cm) long; leaves egg-shaped, have teeth along the margins and develop the characteristic copper coloration.
- Reproduces by seed.

Use and Potential Problem
- Weed, no commercial value.
- Found in wasteland, grassland, and cultivated fields.

Toxicity
- None recorded.

Reference
- Virginia copperleaf [47,48,66,72,74,76]

Woodsorrel, Yellow—*Oxalis stricta* L.

Yellow woodsorrel seeds

Yellow woodsorrel leaf

Yellow woodsorrel seedpod

Yellow woodsorrel flowers

Seed
- Size: 1.3−1.4 mm long; 0.8−0.9 mm wide
- Shape and texture: elliptical with one pointed end, compressed, hilum at the pointed end; surface ridged, dull, about nine sharp crosswise ridges on each surface; ridges bent or interrupted by 2−3 lengthwise wrinkles, under high magnification (10×) ridges are shiny and the interspaces are dull
- Color: dark orange brown

Yellow woodsorrel plant

Leaf shape and arrangement—Leaves: Alternate, with long petioles, have three-notched, heart-shaped leaves, leaves fold up midday and night, smooth leaf surface and margins are often fringed with hairs, pale to grayish green.

Stolon/Rhizome/Roots: No stolons, long, succulent, white to ping rhizomes; stems green to purplish, erect to freely branching and hairy; roots at the lower nodes.

Inflorescence: Flowers in small clusters arising from long stalk at the leaf axils, each flower has 5-sepals, 5 yellow petals, 10 stamens (in two groups of 5), and 5 styles; fruits 5-ridged, cylindrical pointed, erect, hairy capsules containing several brown, ridged seeds, fruit stalks bend backward when matures and the okra-like fruit remains erect.

Woodsorrel, yellow—*Oxalis stricta* L.

Other common names: Pickles, sourgrass, sour clover, creeping lady's sorrel
Family: **Oxalidaceae**
Life cycle: Cool-season, perennial
Native to: North America

Distribution and Adaptation
- Found throughout the United States.
- Found roadsides, turf, gardens, barnyards pastures, dry gravely or stony fields, waste places, fallow, and cultivated field.
- Strives on nutrient-rich moist soils but tolerates a wide range of soil types and environmental conditions.

Morphology/Growth Pattern
- This clover-like plant can be erect or sometimes rooting at the lower nodes without the root stalks, often form a mat, and hairy, leaves alternate, pale green with stout petioles and firm oblong stipules; leaflets in group of three, notched at the tips, flowers yellow; roots at the lower nodes and plant height varies from 1.2 to 20 in. (3—50 cm).
- Reproduces primarily by seed, also can reproduce by rhizomes.

Use and Potential Problem
- No commercial value, considered weed where found.

Toxicity
- Oxalate.

Similar Species
- Yellow woodsorrel is often confused with trifoliate legumes such as clovers and black medic, however, yellow woodsorrel lacks stipules at the base of the petiole and unlike the clovers has heart-shaped leaflets. Also often confused with *creeping woodsorrel (Oxalis corniculata* L.), creeping woodsorrel is more prostrate, frequently roots at the nodes does not have rhizomes but unlike yellow woodsorrel spreads by stolons.

Reference
- Yellow woodsorrel [3,12,14,17,24,33,44,47,48,61,72,76]

Yarrow, Common—*Achillea millefolium* L.

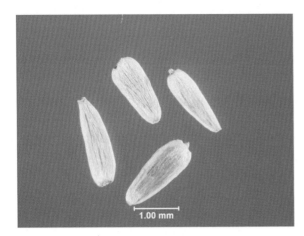

Common yarrow seed

Common yarrow leaves and stem

Common yarrow flower

Seed
- Size: 1.6−1.8 mm long; 0.4−0.5 mm wide
- Shape and texture: long, obovate, small around rim around the basal attachment point, short apical collar and short beak, no pappus; surface membranous, slightly striate, slightly sheen
- Color: dark brown

Common yarrow plant

Leaf shape and arrangement—Leaves: Alternate, finely divided into numerous, short, tiny, feathery grayish green segments, fern-like; covered with white hairs; sessile (no petiole), 4−14 cm long; basal leaves are feather-like and numerous; leaf size reduced upward.

Stolon/Rhizome/Roots: No stolons; extensive rhizomes; stem wooly hairy; fibrous root systems.

Inflorescence: Flowers are in dense flat topped terminal clusters, flower heads are about 6−30 cm across each with 5 white, ray-florets and 10−30 yellow disc florets; there are 20 floral bracts below the flowers appearing in 3−4 overlapping rows; each bract is 3−4 mm long with pale to dark margins; fruits small, oblong, whitish gray.

Yarrow, common—*Achillea millefolium* L.

Other common names: Thousand-leaf, milfoil, bloodwort, old man's pepper
Family: Asteraceae
Life cycle: Cool-season, perennial
Native to: North America and Europe

Distribution and Adaptation
- Found throughout the United States except southwestern states.
- Tolerates many soil types and often found in areas where other plants do not grow such as poor, dry, and sandy soils. Due to its extensive rhizome with fibrous root systems, it tolerates droughty conditions.
- Found in pastures, waste places, fields, and roadsides.

Morphology/Growth Pattern
- Young plant rosette; leaves finely serrated, aromatic when crushed; stem simple to slightly branched, hairy; flower heads in dense flat clusters, white occasionally pink to purple with yellow centers. An extensively branched rhizome allows the plant to form large colonies; grows up to 3 ft (1 m) tall.
- Reproduces by seed and rhizome.

Use and Potential Problem
- No commercial values; seed of common yarrow commonly found in timothy and redtop seed.
- Dairy cattle grazing yarrow can produce off-flavored milk.

Toxicity
- Achillein and astringent.

Similar Species
- Sometimes mistaken for a *fern* (*Athyrium filix-femina*). Also due to finely serrated leaves common yarrow can be confused with Mayweed chamomile, corn chamomile, pineapple-weed, and wild carrot. However, the leaves of these plants are not as finely dissected as mature common yarrow. In addition, unlike common yarrow, those plants do not have rhizomes.

Reference
- Common yarrow [3,12,17,24,33,47,48,61,67,72,76]

Appendix A

Do You Know??

FORBS

Basil, Wild—*Clinopodium vulgare (Satureja vulgaris)*

- Infusion of leaves makes a pleasant cordial tea. Helps overcome stomach weakness.
- European settlers cultivated plants for medicinal purposes.

Buttercup, Bulbous—*Ranunculus bulbosus* L., **Buttercup, Tall Field**—*Ranunculus acris* L., **Buttercup, Creeping**—*Ranunculus repens* L.

- Buttercups contain a glycoside, ranunculin. When plant is crushed it produces yellow volatile oil, protoanemoni, which is unstable. The dry plant (hay) is nontoxic.
- Protoanemoni is an irritant and may cause blisters on the lips and irritations of the mouth and digestive system.

Chickweed, Common—*Stellaria media* L., **Chickweed, Mouse-ear**—*Cerastium vulgatum* L.

- Common chickweed is used in salads; mouse-ear chickweed is too coarse to use raw. Need to be cooked and tastes similar to spinach. Included in soups and stews. All parts of the plant (stems and leaves) are edible.
- Chickens graze chickweed (where it gets its name).
- Has a long history of herbal use in the external treatment of any kind of itching skin condition.

Clover, White—*Trifolium repens* L.

- White clover blossoms were used in folk medicine for treatment of gout, rheumatism, and leucorrhea.
- Some believed that the texture of fingernails and toenails would improve after drinking clover blossom tea.
- The fresh leaves and flowers can be used in salads; flower heads can be dried and used to make tea.
- White clover is thought to clean the system, decreasing irritation and muscular activity of the gastrointestinal tract.

Clover, Sweet—*Melilotus officinalis* Lam.; **Yellow Sweet Clover**—*Melilotus alba* Medik—white sweet clover

- Coumarin is an aromatic compound that is found in many other plants and is used in perfumes as a flavoring agent.
- Honey bees transform the nectar from the white and yellow flowers into a supremely unique, light honey.
- If you let the leaves dry and crush them, they smell like vanilla and make a good sachet.

Dandelion—*Taraxacum officinale* G.H. Weber ex Wiggers

- Traditionally dandelion has been used as a remedy for jaundice and other disorders of the liver and gallbladder, and as a remedy for counteracting water retention.
- The roots of the plant have the most activity regarding the liver and gallbladder, while the diuretic activity is specific to the leaves.

Dock, Curly—*Rumex crispus* L.

- The leaves are very rich in vitamins A and C and can be eaten raw or cooked. Leaves can be added to salads or to soups, only a few leaves should be used. In early spring and fall, can often be pleasant to taste. Oxalic acid content will be reduced if cooked and is said to be safe for consumption if used in a small quantity.
- The roasted seed has been used as a coffee substitute.
- Important source of food for the caterpillars of many butterflies.

Henbit—*Lamium amplexicaule* L.

- All parts of red henbit are used in salads.
- The leaves used as a tea to reduce both internal and external bleeding, and to relieve diarrhea.
- Also used in flower arrangements because of its unusual leaf shape and arrangement.

Horsenettle—*Solanum carolinense* L.

- Many years ago its fruits were used as sedatives and antispasmodics to treat epilepsy.
- The Cherokee would use crushed horsenettle leaves in milk to kill flies.

Horseweed—*Conyza canadensis* (L.) Cronq.

- American Indians made tea from the root and lower stalks to treat lower abdominal pain.

Jimsonweed—*Datura stramonium* L., Synonyms: *Datura tatula* L.

- Jimsonweed has been used by Native Americans and others for drug-induced ceremonial and spiritual purposes.
- Jimsonweed is also called Jamestown weed for two reasons: for the town in Virginia where jimsonweed is believed to have been imported to the United States from England. In 1676 a massive poisoning of soldiers (by eating the plant in salads) in Jamestown, VA occurred, giving rise to the common name "Jamestown weed" and "jimsonweed."
- The seeds and leaves are deliberately used to induce intoxication.
- Atropine, the substance in jimsonweed, has been used in treating Parkinson's disease, peptic ulcers, diarrhea, and bronchial asthma.
- In 1968 the use of jimsonweed as a hallucinogenic drug prompted the US government to ban over-the-counter sales of products prepared from it.

Kudzu—*Pueraria lobata* (Willd.) Ohwi {PUELO} also called *Pueraria thunbergiana*

- The rubber-like vines used to make baskets.
- Used to make paper.
- Used for food such as deep-fried kudzu leaves, kudzu quiche, and many other kudzu dishes.
- In China and Japan, ground kudzu root has been a common ingredient in foods and medications for centuries.
- In China, the kudzu root is used to sober the drunk.

Lamb's-quarters, Common—*Chenopodium album* L.

- Common lamb's-quarters leaves are one source of ascaridole, an oil used to treat round worms and hook worms.
- Used to treat various symptoms attributable to nutritional deficiencies.
- Ground seeds of common lamb's-quarters were used to make bread, cakes, and gruel, and are still used by some American Indians.

Sericea Lespedeza—*Lespedeza cuneata* {Dum. Cours.} G. Don

- Sericea lespedeza is being evaluated as a potential energy crop.

Oxeye Daisy—*Chrysanthemum leucanthemum*

- Besides being planted for its beauty, oxeye daisy was cultivated for home remedies to cure whooping cough, asthma, and other coughs. A tea made from its leaves was sometimes used as an antispasmodic. Today many people enjoy the young leaves of oxeye daisy in salads.

Spiny Sowthistle—*Sonchus asper*

- *Sonchus*, the name given sow thistle by the Romans, is Latin for "hollow" referring to the plant's hollow stems.
- Perennial sowthistle is said to be a favorite of rabbits, hence the many common folk names referring to hares. It was believed predators could not disturb a rabbit sitting beneath the plant.

St. John's Wort—*Hypericum perforatum* L.

- Native Americans used St. John's wort to treat fevers, coughs, intestinal problems, nosebleeds, and snakebites. Powdered preparations of the plant are sold as a dietary supplement which is claimed to relieve symptoms of depression.

Common Teasel—*Dipsacus fullonum* L.

- *Dipsacus* was derived from the Greek verb meaning "to be thirsty," which is likely in reference to the water-collecting cup formed by the stem leaves.
- The plant's common name refers to the practice of using the flowers to tease wool. The common names "card teasel" and "card-thistle" are in reference to the wire brush or "card" used to tease wool.
- Teasel is often added as a dried plant to ornamental arrangements.

Jewelweed—*Impatiens capensis* Meerb.

- The leaf surface of jewelweed is waterproof, i.e., water does not stick to the surface.
- The plant juice is used to reduce inflammation from poison ivy and insect bites.
- Also known to be used by Native Americans to treat stomach cramps.
- Plant juice also used to make dye.

GRASSES

Big Bluestem—*Andropogon gerardii* {L.} Vitman

- Early settlers used big bluestem to build sod homes. Medically, big bluestem is a diuretic and analgesic. The Chippewa used the root for stomach aches.
- On August 31, 1989, the big bluestem was chosen as Illinois' official state prairie grass.

Oats, Cereal—*Avina sativa* L.

- Oats are believed to act as a nervine tonic during periods of stress.
- Cooked oatmeal is used as a poultice to soothe inflammation of the skin, and is a familiar component of cosmetics and herbal bath preparations.
- Also used to relieve diarrhea.

Quackgrass—*Agropyron repens* (L.)

- Quackgrass has been found to have insecticidal properties against mosquito larvae and mollusks, particularly slugs.
- Ground dried rhizomes are used in teas or as a flour source. The grain is also edible but must be parched to remove the chaff but since quackgrass is very subject to ergot care must be taken in using the seed.
- Rhizomes of quackgrass produce a toxic substance commonly known as an allelopathic effect or "chemical warfare" that suppresses the growth of plants growing in the vicinity of quackgrass.

Sweet Vernalgrass—*Anthoxanthum odoratum* L.

- The chemical substance coumarin produced by sweet vernalgrass is used medicinally and also in rat poisons.
- When the dried plant is used internally, coumarin can act to prevent the blood from coaggulating.
- Although the plant is known to cause hay fever, a small amount of its flower has been documented to give immediate relief from hay fever.
- The seed is used for tea.
- The aromatic leaves and dried flowers are used as a strewing herb.
- Leaves are woven into baskets and used in pot-pourri.

Appendix B

The Nutritive Value of Common Pasture Weeds and Their Relation to Livestock Nutrient Requirements

Weeds constantly invade crop fields and pastures; therefore it is important to know the potential quality of individual weed species in making management decisions concerning weed control. It is frequently assumed that weeds have low nutritive value and livestock will not eat weeds, so expensive and time-consuming measures are often used for their control. Some weeds are toxic or poisonous to livestock, and certain weeds are unpalatable—causing a reduction in total intake. Several weed species have thorns or spines that can injure the grazing animal's mouth and/or irritate its eyes, which may lead to pinkeye. Other weeds can cause the milk and meat of livestock to have a negative taste or odor. Weeds also compete with cultivated crops and forages for moisture, light, and nutrients, but many weeds are nutrient-rich and digestible (Tables B.1–B.7).

TABLE B.1 In Vitro Digestible Dry Matter Concentration of Two Perennial Forages and Nine Weeds

Species	1981				1982		
	May 18	June 1	June 15	June 29	May 19	June 1	July 27
				%			
Alfalfa	79[a]	69[b]	62[c]	49[d]	78[e]	68[a]	69[c]
Smooth bromegrass	78[f]	66[g]	59[h]	57[i]	76[j]	67[k]	68[e]
Quackgrass	78[j]	69[f]	63[g]	59[h]	75[j]	69[f]	63[e]
Dandelion	82[c]	84[l]	–	–	78[c]	77[l]	74[e]
White campion	–	–	–	61[m]	80[e]	75[b]	67[c]
Swamp smartweed	–	–	54[e]	49[e]	–	58[e]	34[e]
Perennial sowthistle	–	–	82[b]	66[c]	–	79[e]	–
Jerusalem artichoke	86[e]	81[e]	70[e]	66[e]	81[e]	81[e]	71[e]
Curly dock	–	–	–	–	77[e]	64[e]	50[e]
Hoary alyssum	89[a]	76[c]	64[d]	58[l]	–	–	–
Canada thistle	–	–	76[n]	64[c]	79[e]	78[e]	72[o]

[a]Early bud; [b]Late bud; [c]Mid-bloom; [d]Full bloom; [e]Vegetative; [f]Boot; [g]Head; [h]Anthesis; [i]Green seed; [j]Joint; [k]Early head; [l]Seed; [m]Early bloom; [n]Bud.
Adopted from G.C. Marten, C.C. Sheaffer, D.L. Wyse, Forage nutritive value and palatability of perennial weeds. Agron. J. 79 (1987) 980–986.

TABLE B.2 Quality of Alfalfa Occurring in a Newly Established Stand Compared to Seven Annual Weeds Occurring in a Weed Nursery on July 16, 1971

Species	IVDDM[a]	ADF[b]	CP[c]
		%	
Alfalfa[d]	72	24	27
Redroot pigweed[e]	73	21	25
Common lamb's-quarters	68	22	25
Common ragweed	73	25	25
Pennsylvania smartweed	51	22	24
Yellow foxtail	69	30	20
Giant foxtail	62	33	18
Barnyardgrass	70	33	18

[a]Invitro digestible dry matter; [b]Acid detergent fiber; [c]Crude protein; [d]Alfalfa was seeded on May 14, 1971; [e]Weed nursery was seeded naturally in late summer and autumn of 1970.
Adapted from G.C. Marten, R.N. Andersen, Forage nutritive value and palatability of 12 common annual weeds. Crop Sci. 15 (1975) 821–827.

TABLE B.3 Crude Protein (CP) and In Vitro Dry Matter Digestibility (IVDMD) of Cool Season Weeds and Forages at Three Stages of Maturity

Weeds	Vegetative		Flower/Boot		Fruit/Head	
	CP (%)	IVDMD (%)	CP (%)	IVDMD (%)	CP (%)	IVDMD (%)
Forbs						
Carolina geranium	19	78	19	70	11	68
Curly dock	30	73	19	54	16	51
Cutleaf evening primrose	20	72	14	69	11	52
Henbit	–	–	20	78	16	75
Virginia pepperweed	32	86	26	72	17	63
Grasses						
Cheat	23	81	18	69	14	61
Little barley	24	82	18	78	14	62
Virginia wildrye	23	80	19	74	7	60
Wild oats	23	75	–	–	–	–
Forages						
Hairy vetch	30	80	29	77	26	77
Ladino clover	27	81	22	85	23	83
Rye	28	79	24	81	13	70
Tall fescue	22	78	17	73	13	67

Adapted from S.C. Bosworth, C.S. Hoveland, G.A. Buchanan, Forage quality of selected cool-season weed species. Weed Sci. 34 (1985) 150–154; J.T. Green, Pasture Weeds and Forage, 1998.

TABLE B.4 Crude Protein (CP) and In Vitro Dry Matter Digestibility (IVDMD) of Warm Season Weeds and Forages at Three Stages of Maturity

Weeds	Vegetative		Flower/Boot		Fruit/Head	
	CP (%)	IVDMD (%)	CP (%)	IVDMD (%)	CP (%)	IVDMD (%)
Forbs						
Bur gherkin	–	–	17	75	14	79
Coffee senna	17	81	22	75	15	67
Common purslane	–	–	19	80	–	–
Cypressvine morningglory	20	80	–	–	13	77
Florida beggarweed	22	74	17	65	13	55
Hemp sesbania	31	70	14	66	11	52
Ivyleaf morningglory	20	80	–	–	11	78
Jimsonweed	25	72	21	66	17	59
Prickly sida	17	80	18	70	12	56
Redroot pigweed	24	73	17	71	11	64
Sicklepod	22	84	14	76	17	71
Tall morningglory	20	82	–	–	14	76
Grasses						
Crabgrass	14	79	8	72	6	63
Crowfootgrass	16	67	8	54	9	43
Fall panicum	19	72	9	63	7	54
Texas panicum	16	74	11	62	8	52
Yellow foxtail	18	73	12	66	14	57
Forages						
Bermudagrass	16	58	7	51	8	43
Pearlmillet	17	59	6	60	8	60

Adapted from S.C. Bosworth, C.S. Hoveland, G.A. Buchanan, W.B. Anthony, Forage quality of selected warm-season weed species. Agron. J. 72 (1980), 1050–1054; J.T. Green, Pasture Weeds and Forage, 1998.

TABLE B.5 Neutral Detergent Fiber Concentration of Two Perennial Forages and Nine Weeds

Species	1981				1982		
	May 18	June 1	June 15	June 29	May 19	June 1	July 27
				%			
Alfalfa	31[a]	45	51	64	30	42	35
Smooth bromegrass	49	65	67	66	47	63	56
Quackgrass	46	59	66	64	41	53	56

(Continued)

TABLE B.5 (Continued)

Species	1981				1982		
	May 18	June 1	June 15	June 29	May 19	June 1	July 27
				%			
Dandelion	26	30	–	–	27	33	25
White campion	–	–	–	58	35	46	48
Swamp smartweed	–	–	44	44	–	35	40
Perennial sowthistle	–	–	31	45	–	27	–
Jerusalem artichoke	22	34	47	49	24	29	32
Curly dock	–	–	–	–	24	33	33
Hoary alyssum	29	42	52	60	–	–	–
Canada thistle	–	–	41	50	28	32	34

[a]Same maturity stages as indicated for each species and date in Table B.1.
Adapted from G.C. Marten, C.C. Sheaffer, D.L. Wyse, Forage nutritive value and palatability of perennial weeds. Agron. J. 79 (1987) 980–986.

TABLE B.6 Quality at Two Growth Stages of Weeds in Spring-Sown Alfalfa

Weed	Harvest Date in July	Growth Stage	CP	IVDDM	NDF	ADF
				%		
Common lamb's-quarters	2	Bud	22	73	22	17
Common lamb's-quarters	7	Flower	18	67	27	19
Shepherds purse	2	Green seed	19	55	37	29
Shepherds purse	7	Seed	16	53	41	34
Pennsylvania smartweed	2	Flower	18	47	24	19
Pennsylvania smartweed	7	Late flower	15	44	32	19
Redroot pigweed	2	Flower	18	74	22	16
Redroot pigweed	7	Early seed	15	73	27	20
Yellow foxtail	2	Early seed	17	63	52	27
Yellow foxtail	7	Seed	14	60	54	30
Common ragweed	2	Veg.[a]	26	77	21	17
Common ragweed	7	Veg.	21	70	26	21
Alfalfa	7	Early bloom	20	70	28	23

[a]Vegetative.
Adapted from D.G. Temme, R.G. Harvey, R.S. Fawcett, A.W. Young, Effects of annual weed control on alfalfa forage quality. Agron. J. 71 (1979) 51–54.

TABLE B.7 Crude Protein Concentration of Two Perennial Forages and Nine Weeds

Species	1981				1982		
	May 18	June 1	June 15	June 29	May 19	June 1	July 27
				%			
Alfalfa	27[a]	20	15	14	26	20	21
Smooth bromegrass	16	11	8	7	23	14	18
Quackgrass	17	13	9	7	27	18	19
Dandelion	17	12	–	–	20	13	20
White campion	–	–	–	11	26	15	14
Swamp smartweed	–	–	17	14	–	22	17
Perennial sowthistle	–	–	16	13	–	21	–
Jerusalem artichoke	27	18	11	10	29	19	22
Curly dock	–	–	–	–	28	17	20
Hoary alyssum	20	14	12	7	–	–	–
Canada thistle	–	–	17	15	28	19	18

[a]Same maturity stages as indicated for each species and date in Table B.1.
Adapted from G.C. Marten, C.C. Sheaffer, D.L. Wyse, Forage nutritive value and palatability of perennial weeds. Agron. J. 79 (1987) 980–986.

FURTHER READING

[1]   D.M. Ball, G.D. Lacefield, N.P. Martin, D.A. Mertens, K.E. Olson, D.H. Putnam, et al., Understanding Forage Quality, American Farm Bureau Federation Publication 1-01, Park Ridge, IL, 2001.

[2]   S.C. Bosworth, C.S. Hoveland, G.A. Buchanan, Forage quality of selected cool-season weed species, Weed Sci. 43 (1985) 150–154.

[3]   S.C. Bosworth, C.S. Hoveland, G.A. Buchanan, W.B. Anthony, Forage quality of selected warm-season weed species, Agron. J. 72 (1980) 1050–1054.

[4]   W.S. Curran, D.D. Lingenfelter, Weed Management in Pasture Systems, Penn. State Univ. Pub. CAT UC172, University Park, PA, 2001.

[5]   S.L. De Bruijn, E.W. Bork, Biological control of Canada thistle in temperate pastures using high density rotational cattle grazing, Biol. Control 36 (2006) 305–315.

[6]   T.E. Dutt, R.G. Harvey, R.S. Fawcett, Feed quality of hay containing perennial broadleaf weeds, Agron. J. 74 (1982) 673–676.

[7]   J.B. Hall, W.W. Seay, S.M. Baker, Nutrition and Feeding of the Cow-Calf Herd: Production Cycle Nutrition and Nutrient Requirements of Cows, Pregnant Heifers and Bulls, Extension Publication 400-012. Virginia Cooperative Extension, Blacksburg, VA, 2001. <http://www.ext.vt.edu/pubs/beef/400-012/400-012.html>.

[8]   P.J. Holst, C.J. Allan, M.H. Campbell, A.R. Gilmour, Grazing of pasture weeds by goats and sheep. 1. Nodding thistle (Carduus nutans), Aust. J. Exp. Agric. 44 (2004) 547–551.

[9]   W.M. Lewis, J.T. Green Jr., Weed management, Production and Utilization of Pastures and Forages, Technical Bulletin 305. North Carolina Agricultural Research Service, North Carolina State University, Raleigh, NC, 1995.

[10] J.-M. Luginbuhl, J.T. Green, J.P. Mueller, M.H. Poore, Forage needs for meat goats and sheep, Production and Utilization of Pastures and Forages, Technical Bulletin 305. North Carolina Agricultural Research Service, North Carolina State University, Raleigh, NC, 1995.

[11] J.-M. Luginbuhl, T.E. Harvey, J.T. Green Jr., M.H. Poore, J.P. Mueller, Use of goats as biological agents for the renovation of pastures in the Appalachian region of the United States, Agroforest Syst 44 (1999) 241–252.

[12] G.C. Marten, R.N. Andersen, Forage nutritive value and palatability of 12 common annual weeds, Crop Sci. 15 (1975) 821–827.

[13] G.C. Marten, C.C. Sheaffer, D.L. Wyse, Forage nutritive value and palatability of perennial weeds, Agron. J. 79 (1987) 980–986.

[14] National Research Council (NRC), Nutrient Requirements of Beef Cattle, seventh ed., The National Academies Press, Washington, DC, 1996.

[15] B.M. Santos, J.A. Dusky, T.A. Bewick, D.G. Shilling, J.P. Gilreath, Phosphorus absorption in lettuce, smooth pigweed (Amaranthus hybridus), and common purslane (Portulaca oleracea) mixtures, Weed Sci. 52 (2004) 389–394.

[16] K. Sedivec, T. Hanson, C. Heiser, Controlling Leafy Spurge Using Goats and Sheep, Extension Publication R-1093. North Dakota Extension, Fargo, ND, 1995.

[17] C.C. Stallings, Updated Nutrient Specifications for the Dairy Herd, Extension Publication 404-105. Virginia Cooperative Extension, Blacksburg, VA, 1996. <http://www.ext.vt.edu/pubs/dairy/404-105/404-105.html>.

[18] D.G. Temme, R.G. Harvey, R.S. Fawcett, A.W. Young, Effects of annual weed control on alfalfa forage quality, Agron. J. 71 (1979) 51−54.

[19] M. Vengris, M. Drake, W.G. Colby, J. Bart, Chemical composition of weeds and accompanying crop plants, Agron. J. 45 (1953) 213−218.

Bibliography

[1] W.P. Anderson, Perennial Weeds: Characteristics and Identification of Selected Herbaceous Species, Iowa State Univ. Press, Ames, IA, 1999.

[2] S.G. Archer, C. Bunch, The American Grass Book, University of Oklahoma Press, USA, 1953.

[3] F.L. Baldwin, E.B. Smith, Weeds of Arkansas. Lawns, Turf, Roadsides, Recreation Areas. A Guide to Identification, Coop. Ext. service Univ. of Arkansas Division of Agriculture, United State Dept. Agric, and County Governments Cooperating. Printed by the Univ. Arkansas Printing Service, Fayetteville, AK, 1981.

[4] D.M. Ball, C.S. Hoveland, G.D. Lacefield, Southern Forages, Potash and Phosphate Institute, Atlanta, GA, 2002.

[5] D. Ball, Forage Chicory, Dept. of Agronomy & Soils, Auburn University, Auburn, AL, 1997.

[6] T.M. Barkley, R.E. Brooks, E.K. Schofield (Eds.), Flora of the Great Plains, University Press of Kansas, Lawrence, KS, 1986.

[7] R.F. Barnes, D.A. Miller, C.J. Nelson, Forages. Volume 1. An Introduction to Grassland Agriculture, Iowa State University Press, Ames, IA, 1995.

[8] H.L. Blomquist, The Grasses of North Carolina, Duke University Press, Durham, NC., 1948.

[9] S.C. Bosworth, C.S. Hoveland, G.A. Buchanan, W.B. Anthony, Forage quality of selected warm-season weed species, Agron. J. 72 (1980) 1050–1054.

[10] G.W. Burton, W.G. Monson, Inheritance of dry matter digestibility in bermudagrass, *Cynodon dactylon* (L.) Pers, Crop Sci. 12 (1972) 375–378.

[11] C.G. Chambliss, D.S. Wofford, White Clover, Florida Cooperative Extension. Document AA198. Institute of Food and Agricultural Sciences, University of Florida, 2003.

[12] P.R. Cheeke, Natural Toxicants in Feeds, Forages, and Poisonous Plants, Interstate Publishers, Inc., Danville, IL, 1998.

[13] N. Christians, Fundamentals of Turfgrass Management, second ed., John Wiley and Sons, Inc, Hoboken, NJ, 2004, pp. 61–63.

[14] Colvin, D.L., R. Dickens, J.W. Everest, D. Hall, and L.B. McCarty. Weeds of Southern Turfgrasses. Golf Courses, Lawns, Roadsides, Recreational Areas, Commercial Sod. Cooperative Extension Service, University of Georgia College of Agricultural and Environmental Sciences. Athens, GA.

[15] D.S. Chamblee, J.T. Green, Jr. (Eds.), Production and Utilization of Pastures and Forages in North Carolina, North Carolina State University Agricultural Research Service, 1995, Tech. Bull., 305, 162 pp.

[16] W. Darlington, American Weeds and Useful Plants, Orange Judd, Publishers, New York, NY, 1859.

[17] L.W. Davis, Weed Seeds of the Great Plains: A Handbook for Identification, Cooperative Extension Service of Kansas State University, University Press of Kansas, 1993.

[18] R.J. Delorit, C.R. Gunn, Seeds of Continental United States Legumes (Fabaceae), F. A. Weber and Sons, Lithographers, Park Falls, WI, 1986.

[19] R.J. Delorit, An Illustrated Taxonomy Manual of Weed Seeds, Agronomy Publications, River Falls, WI, 1970.

[20] M.L. Fernald, A.C. Kinsey, Edible Wild Plants of Eastern North America, Harper Brothers, New York, NY, 1951.

[21] A. Georgia, Manual of Weeds, Macmillan Publishers, New York, NY, 1914.

[22] P.B. Gibson, W.A. Cope, in: N.L. Taylor (Ed.), Clover Science and Technology, ASA, CSSA, SSSA, Inc., Madison, WI, 1985.

[23] J.M. Gillett, N.L. Taylor, in: M. Collins (Ed.), The World of Clovers, Iowa State University Press, Ames, IA, 2001.

[24] P.D. Haragan, Weeds of Kentucky and Adjacent States: Field Guide, University Press of Kentucky, Lexington, KY, 1953.

[25] J.W. Hardin, Stock-Poisoning Plants of North Carolina, Bulletin No. 414. Agric. Exp. Stn. North Carolina State Univ., Raleigh, NC, 1966.

[26] H.D. Harrington, Y. Matsumura, Edible Native Plants of the Rocky Mountains, University of New Mexico Press, Albuquerque, NM, 1967.

[27] E. Hafliger, H. Scholz, Grass Weed 2, Ciba-Geigy, Basel, 1981.

[28] A.W. Hatfield, How to Enjoy Your Weeds, Frederick Muller, London, 1969.

[29] M.E. Heath, R.F. Barnes, D.S. Metcalfe, Forages: The Science of Grassland Agriculture, fourth ed., Iowa State University Press, Ames, IA, 1985.

[30] M.E. Heath, D.S. Metcalfe, R.F. Barnes, Forages: The Science of Grassland Agriculture, third ed., The Iowa State University Press, Ames, IA, 1973.

[31] H.D. Hughes, M.E. Heath, D.S. Metcalfe, Forages, Iowa State University Press, Ames, IA, 1962.

[32] A.S. Hitchcock, Manual of the Grasses of the United States, second ed., Misc. publ. no. 200. USDA, Washington, DC, 1950.

[33] D. Isely, Weed Identification and Control. In the North Central States, Iowa State University press, Ames, IA, 1960.

[34] J.R. Johnson, G.E. Larson, Grassland Plants of South Dakota and the Northern Great Plains, South Dakota State University. College of Agriculture and Biological Sciences, South Dakota Agric. Exp. Stn. Brookings, South Dakota, 1999.

[35] G.A. Jung, J.A. Shaffer, W.L. Stout, Switchgrass and big bluestem responses to amendments on strongly acid soils, Agron. J. 80 (1988) 669–676.

[36] R.C. Kinch, L. Wrage, R.A. Moore, South Dakota Weeds, South Dakota State Weed Commission, South Dakota, 1975.

[37]   J.M. Kingsbury, Poisonous Plants of the Unites States and Canada, Prentice-Hall Inc., Englewood Cliffs, NJ, 1964.

[38]   G.E. Larson, J.R. Johnson, Plants of the Black Hills and Bear Lodge Mountains, South Dakota State Univ. South Dakota Agric. Exp. Stn. Brooking, South Dakota, 1999, pp. 230–231.

[39]   G.E. Larson, J.R. Johnson, Plants of the Black Hills and Bear Lodge Mountains, Fenske Media Corporation, Rapid City, SD, 1999.

[40]   H.L. Leithead, L.L. Yarlett, T.N. Shiflet, 100 Native Forage Grasses in 11 Southern States, Agricultural Handbook No. 389. Soil Conservation Service. USDA, Washington, DC, 1976.

[41]   Lost Crops of Africa, Board on Science and Technology for International Development, National Academy Press, Vol. I. Grains, 1996.

[42]   A.B. Massey, Farm Weeds: Their Importance and Control, Virginia Cooperative Wildlife Research Unit. Dept. Biology, Virginia Tech, Blacksburg, VA, 1956.

[43]   A.G. Mathes, Management, in: R.C. Buchner, L.P. Bush (Eds.), Tall Fescue, American society of Agronomy, Madison, WI, 1979.

[44]   L.B. McCarty, J.W. Everest, D.W. Hall, T.R. Murphy, F. Yelverton, Color Atlas of Turfgrass Weeds, Sleeping Bear Press, Chelsea, MI, 2001.

[45]   D.A. Miller, Forage Crops, McGraw-Hill Book Co., New York, NY, 1984.

[46]   J.H. Miller, K.V. Miller, Forest Plants of the Southeast and their Wildlife Uses, Southern Weed Science Society, 1999.

[47]   F.H. Montgomery, Weeds of the Northern United States and Canada, Department of Botany, Ontario Agricultural College. Frederick Warne Inc., New York, NY, 1964.

[48]   W.C. Muenscher, Weeds, Comstock Publishing Associates, A Division of Cornell University Press, Ithaca, NY; London, 1980.

[49]   A.F. Musil, Identifications of Crop and Weed Seeds, USDA. Agricultural Marketing Service. Agricultural Handbook No. 219, Washington, DC, 1979.

[50]   Native Grasses, Legumes and Forbs, Section 1 of a series pasture and range plants, Phillips Petroleum Company, 1955.

[51]   F.S. Nowosad, D.E.N. Wales, W.G. Dore, The Identification of Certain Native and Naturalized Hay and Pasture Grasses by their Vegetative Characters, MacDonald College, P. Q., 1942.

[52]   P.D. Ohlenbusch, Range Grasses of Kansas, Kansas State University Cooperative Extension Service Publication C-567, Manhattan, KS, 1976.

[53]   O.M. Scott & Sons, Guide to the Identification of Grasses, O. M. Scott & Sons, Marysville, OH, 1985.

[54]   O.M. Scott & Sons, Guide to the Identification of Dicot Turf Weeds, O. M. Scott & Sons, Marysville, OH, 1985.

[55]   Pasture Legumes and Grasses. A Guide to the Identification and Use of Selected Species for Pasture Improvement, Bank of New South Wales, Sydney, 1965.

[56]   C.E. Phillips, Some Grasses of the Northeast. A Key to their Identification by Vegetative Characters, University of Delaware. Field Manual No. 2 – July, 1962, Newark, DE, 1962.

[57]   Phillips Petroleum Company, Pasture and Range Plants, 1963.

[58]   A.E. Radford, H.E. Ahles, C.R. Bell, Manual of the Vascular Flora of the Carolinas, University of North Carolina Press, Chapel Hill, NC, 1973.

[59]   M. Rasnake, G.D. Lacefield, J.C. Henning, N.L. Taylor, D.C. Ditsch, Growing White Clover in Kentucky, AGR-93, 1993.

[60]   J.R. Rooney, J.L. Robertson, Equine Pathology, Iowa State University Press, Ames, IA, 1996.

[61]   F. Royer, R. Dickinson, Weeds of the Northern U.S. and Canada, Lone Pine Pub. The University of Alberta Press, Alberta, 1960.

[62]   A.B. Russell, J.W. Hardin, L. Grand, A. Fraser, Poisonous Plants of North Carolina, North Carolina State University, Raleigh, NC, 1997. Available from: http://www.ces.ncsu.edu/depts/hort/consumer/poison/Daturst.htm.

[63]   M. Sarrantonio, Rodale Institute. Managing Cover Crops Profitably, Sustainable Agriculture Research and Education Program, USDA, 1994.

[64]   R.G. Smith, Grass, Identification, Texas A&M University Agricultural Research and Extension Center, 1974.

[65]   South Dakota Weeds, Agric. Ext. Serv. South Dakota State University, Pub. South Dakota State Weed Control Commission, 1975.

[66]   Southern Weed Science Society Weed Identification Committee. Weed Identification Guide. No. 5. Southern Weed Science Society, Champaign, IL.

[67]   E.R. Spencer, All About Weeds, Dover Publications, New York, NY, 1968.

[68]   J. Stubbendieck, S.L. Hatch, C.H. Butterfield, North American Range Plants, University of Nebraska Press, Lincoln; London, 1997.

[69]   A.J. Turgeon, Turfgrass Management, Prentice Hall Inc., Upper Saddle River, NJ, 1980.

[70]   A.H. Bruneau (Ed.), Turfgrass Pest Management Manual. A Guide to Major Turfgrass Pests and Turfgrasses, Cooperative Extension Service, AG-348, Raleigh, NC, 2006.

[71]   R.H. Uva, J.C. Neal, J.M. DiTomaso, Weeds of the Northeast, Cornell University Press, Ithaca, NY, 1997.

[72]   F.C. Weintraub, Grasses Introduced into the United States, USDA Forest Service. Agric. Handbook. No. 58, 1958, p. 42.

[73]   T.D. Whitson, L.C. Burrill, S.A. Dewey, D.W. Cudney, B.E. Nelson, R.D. Lee (Eds.), Weeds of the West, fifth ed., Pioneer of Jackson Hole, Jackson, WY, 1996.

[74]   R.L. Wilbur, The Leguminous Plants of North Carolina, North Carolina Agric. Exp. Stn. Number, 1963, pp. 294.

[75]   R.E. Wilkinson, H.E. Jaques, How to Know the Weeds, Wm. C. Brown Company Publishers, USA, 1972.

[76]   C.L. Wilson, W. Loomis, T.A. Steens, Botany, fifth ed., Holt, Rinehart and Winston, Publishers, New York, NY, 1971.

[77]   D. Wolfe, R.S. White, S.E. Tinsley, Establishing and Managing Caucasian Bluestem, Virginia Cooperative Extension. Pub. No. 418-014, 1996.

[78]   R.L. Zimdahl, Weeds and Words, Iowa State University Press, Ames, IA, 1989.

[79]   <http://www.ppws.vt.edu/scott/weed_id/genusindex.htm>.

Index by Scientific Name

Index by Common Name

General Index

Note: Page numbers followed by "*t*" refer to tables.

Printed in the United States
By Bookmasters